Biotechnics
&
Society

Biotechnics
&
Society

The Rise of
Industrial Genetics

Sheldon Krimsky

PRAEGER

New York
Westport, Connecticut
London

Library of Congress Cataloging-in-Publication Data

Krimsky, Sheldon.
    Biotechnics and society : the rise of industrial genetics /
Sheldon Krimsky.
        p.   cm.
    Includes bibliographical references and index.
    ISBN 0–275–93859–X (alk. paper).—ISBN 0–275–93860–3 (pbk. :
alk. paper)
        1. Genetic engineering—Industrial applications—Social aspects.
    2. Biotechnology—Social aspects.   I. Title.
    TP248.6.K75   1991
    303.48'3—dc20   90–23214

British Library Cataloguing in Publication Data is available.

Library of Congress Catalog Card Number: 90–23214
ISBN: 0–275–93859–X
ISBN: 0–275–93860–3 (pbk.)

First published in 1991

Praeger Publishers, One Madison Avenue, New York, NY 10010
An imprint of Greenwood Publishing Group, Inc.

Printed in the United States of America

The paper used in this book complies with the
Permanent Paper Standard issued by the National
Information Standards Organization (Z39.48–1984).

10 9 8 7 6 5 4 3 2

In memory of my father
Alex Krimsky

Contents

Part III: Social Controls

10. The Growing Complexity of Regulation   181

11. Biotechnology Assessment: Dilemmas and Opportunities   205

    Notes   231

    Selected Bibliography   253

    Index   255

Tables and Figures

Tables

Figures

Preface

Technological change is most commonly viewed ex post facto. Decades or centuries after technologies have been adopted, historians try to re-create the decisions that were made, sort out the influencing factors, and debate the options that were available at the time. Studying a technological revolution retrospectively is very different than watching it develop. In the former case we are less mindful of what could have been because the present sets boundaries on our understanding and expectations of the past. As we study a technological revolution prospectively, in the process of unfolding, we may ask: How will it be realized? Will there be any regrets? What are the sources of criticism? How are decisions being made? Is society prepared or capable of anticipating the impacts?

The farther away we are from a critical node of technological innovation, the more faint become the voices of dissent and the advocates of alternative futures. Technological changes become redefined as inevitable outcomes of some inexorable progressive historical movement. David Noble writes in *Forces of Production* that "our culture objectifies technology and sets it apart and above human affairs. Here technology comes to be viewed as an autonomous process, having a life of its own which proceeds automatically, and almost naturally, along a singular path. Supposedly, self-defining and independent of social power and purpose, technology appears to be an external force impinging upon society, as it were from outside, determining events to which people must forever adjust."[1] This book is an antidote to such an acultural view of technological change.

Because we do not know the outcome of emerging innovations in biotechnology, it gives us the latitude to discuss their social significance with a freshness and critical skepticism. It is quite different when we study a mature revolution where the well-established successes and failures influence what is selected for analysis.

My goal in this book is to make available to contemporary readers and to those who will be reexamining the history of biotechnology several generations from now the opportunity to understand what questions were being asked, what options were available, and what decisions were being made when the industrial application of genetic technologies was still in its infancy.

I have been observing the developments in biotechnology since 1975 when the recombinant DNA controversy was just getting underway. By 1980 the commercialization of genetic technologies was expanding rapidly into major industrial sectors. After more than ten years of commercial investment in applied molecular genetics, it is an opportune time to take account of the developments, expectations, and emergent social, political, and ethical issues that mark this phase of the genetics revolution.

The chapters in this work are used both as a window and a lens through which to view the social discourse around biotechnology. The window offers an informational account of key events and decision points, albeit a selected representation of central events. It addresses questions such as: What approach was taken for regulating a new technology? How were different sectors of society and factions within those sectors engaged in the issues? What were the early expectations for the first generation of products and how did they compare with the outcome? The reader is taken through different stages in the process of policy formation.

My analysis of the public dialogue draws upon many diverse sources of information, including media accounts, government documents and memoranda, public interest sources, personal interviews, and scientific and scholarly studies. But these sources themselves are not sufficient to fulfill the book's mission. I have also been a participant observer in the national and international discussions.

I served on the National Institutes of Health's Recombinant DNA Advisory Committee from 1978 to 1981, as an advisor to a presidential commission on bioethics, and as a consultant to the Office of Technology Assessment for its project titled "New Developments in Biotechnology." I have been on study panels of church groups and served as advisor to local municipalities. I played a part in drafting the first legislation passed in the United States (in Cambridge, Massachusetts) on the use of recombinant DNA technology. In addition, I have been a member of the board of directors of the Council for Responsible Genetics—a public interest organization formed to advance public understanding and constructive debate over the applications of biotechnology.

I make no claims to approaching the subject matter as a disinterested party; but I have tried to share my perspective and feeling for the events as they have evolved without compromising the historical record. The participant observer adds something unique to the interpretation of published sources. However, no single observer can claim a monopoly on the definitive interpretation of complex events that surround the birth of an industrial revolution. I hope that this book stimulates other participants to explore this critical historical period from their unique perspectives.

The lenses I offer in the work are the prescriptive frameworks for addressing social, environmental, and ethical impacts of genetic engineering. I have struggled with issues like: What is the responsibility of government in regulating biotechnology? How much effort should be put into evaluating the hypothetical risks of biotechnology? Should universities enter into partnerships with industry to earn profits from faculty research? Should we use genetic engineering on humans? Should animals be patentable subject matter? Is there a public role for directing biotechnology toward specific ends or should the market decide which products get developed? These queries are the basis of the "second generation" debates in biotechnology. On the surface they are about the alteration of biological life forms, but they are also part of a wider dialogue. These debates are an important expression of the prevailing public attitudes toward technology, of scientific-social interactions, and of the means through which our society grapples with scientific uncertainty and technological change. Will biotechnology contribute to a more harmonious and sustainable relationship between humans and nature? Will it help to flatten out the vast disparities in wealth on the global scale? Will it make meaningful contributions to improving health and solving the mystery of great epidemics? These are not simple questions to answer. They will take considerable amount of continued study with adequate funding. Some agencies, both state and federal, are beginning to see the value in setting aside a percentage of science and technology research budgets for the study of the social and ethical impacts of research. This is truly a sign of a more highly realized culture. We have long passed the point where we can assert in an uncritical way the synonymity of science/technology with progress. The critical debates over biotechnological futures represented in this work are, in some degree, a proxy for disagreements about the meaning of progress.

There are many people to whom I am indebted. Most know nothing about the development of this book. But it was their scholarly contributions that provided the direction and inspiration for this work. Sections of this book were supported by conference invitations and grants. The University of New Hampshire's invitation to participate in an international symposium on "Universities in the Twenty-First Century" (1985) was the stimulus for much of the analysis in Chapter 4. I wish to thank James G. Ennis and Robert Weissman, who helped me collect and analyze the data in this chapter, and Mark Rossi, who was a research assistant for the project.

The Center for Environmental Management at Tufts University, formerly headed by Anthony D. Cortese, provided grant support for research into the "ice minus" case presented in Chapter 7. John R. Fowle, III, invited me to contribute a chapter to his book *Application of Biotechnology*.[2] That essay provided the groundwork for Chapters 6 and 8.

I appreciate the support of the Tufts University administration for approving my sabbatic leave and for furnishing me with research assistance. I warmly acknowledge the help of Leah Steinberg and Elaine Geyer-Allely, who checked references, provided bibliographical support, and helped with software problems.

I am indebted to Ruth Hubbard for her insights into the social role of biological knowledge, to the writings of Edward Yoxen, Martin Kenney, Jack Kloppenburg,

Jr., Frederick Buttel, and Marc Lappé who contributed much to our understanding of the early developments of biotechnology, to the works of Calestous Juma, the Rural Advancement Fund International, and the International Coalition for Development Action for their analyses of the international impacts of biotechnology, to the board of directors of the Council for Responsible Genetics for many hours of stimulating discussions and debates about the future directions of biotechnology, and to the many historians and sociologists of technology, particularly J. D. Bernal, Lewis Mumford, and David Noble, who have redefined the study of technology as a social enterprise. Finally, I want to thank my family, Carolyn Boriss-Krimsky, Alyssa, and Eliot for their loving support.

Acronyms

AEC	Atomic Energy Commission
AGS	Advanced Genetic Sciences
AIDS	Acquired Immunodeficiency Syndrome
ATSAC	Administrator's Toxic Substances Advisory Committee
BGH	Bovine Growth Hormone
BSB	Biotechnology Science Board
BSCC	Biotechnology Science Coordinating Committee
BST	Bovine Somatotropin
CCPA	Court of Customs and Patent Appeals
CCT	Concerned Citizens of Tulelake
CRG	Council (formerly Committee) for Responsible Genetics
CRTM	Californians for Responsible Toxics Management
DABS	Dual-affiliated Biotechnology Scientists
DES	Diethylstilbestrol
DNA	Deoxyribonucleic Acid
DOD	Department of Defense
DOE	Department of Energy
EIR	Environmental Impact Review
ELI	Environmental Law Institute
EOP	Executive Office of the President
EPA	Environmental Protection Agency

ESA	Ecological Society of America
EUP	Experimental Use Permit
FCCSET	Federal Coordinating Council on Science, Engineering and Technology
FET	Foundation on Economic Trends
FIFRA	Federal Insecticide, Fungicide and Rodenticide Act
GEM	Genetically Engineered Microorganism
GEO	Genetically Engineered Organism
HED	Hazards Evaluation Division
HEW	Health Education and Welfare (Department of)
HGE	Human Genetic Engineering
HGH	Human Growth Hormone
HGT	Human Gene Therapy
HHS	Health and Human Services (Department of)
HSCT	Human Somatic Cell Therapy
IBC	Institutional Biosafety Committee
INA	Ice Nucleating Agent
IRB	Institutional Review Board
LEC	Large Established Companies
MEOR	Microbially Enhanced Oil Recovery
NAS	National Academy of Sciences
NBE	New Biotechnology Enterprise
NBF	New Biotechnology Firm
NIH	National Institutes of Health
NRDC	Natural Resources Defense Council
NSB	National Science Board
NSF	National Science Foundation
OPTI	Office of Productivity, Technology and Innovation
OPTS	Office of Pesticides and Toxic Substances
OSHA	Occupational Safety and Health Administration
OSTP	Office of Science and Technology Policy
OTA	Office of Technology Assessment
OTS	Office of Toxic Substances
PCB	Polychlorinated Biphenyl

PMN	Premanufacture Notice
PSAC	President's Science Advisory Committee
RAC	Recombinant DNA Advisory Committee
RAFI	Rural Advancement Fund International
RCRA	Resource Conservation Recovery Act
RDLP	Research Development Limited Partnership
rDNA	Recombinant DNA
SAB	Scientific Advisory Board
SCID	Severe Combined Immune Deficiency
TSCA	Toxic Substances Control Act
TPA	Tissue Plasminogen Activator
UIRR	University Industry Research Relationship
USDA	U.S. Department of Agriculture
RDLP	Research and Development Limited Partnership

Biotechnics
&
Society

1

The Cultural Significance of the Genetics Revolution

And we make by art, in the same orchards and gardens, trees and flowers, to come earlier or later than their seasons, and to come up and bear more speedily than by their natural course they do.

Francis Bacon, 1622[1]

A technological assault is being prepared that will transform the economies of developed and developing nations. Its substance is the engineering of life processes for commercial ends.

Edward Yoxen, 1983[2]

A BIOTECHNOLOGICAL REVOLUTION

The impact of technological innovation may be measured by the depth and breadth of its penetration into the industrial economy. The depth of its penetration is determined by how great a transformation the technology makes within a single sector. Thus, if advances in photovoltaic technology were to result in solar cells that generate 50 percent of the domestic electricity, this advance would be considered a "deep" revolution in energy production. The breadth of a technological revolution signifies how many sectors of the economy are affected. In 1908, L. H. Baekeland invented a synthetic substance called bakelite that ushered in the age of plastics. Advances in material sciences such as this have affected every sector of the industrial and consumer economy and as such represent a technological revolution of considerable penetration.

By most accounts, several discoveries that were made in the biomedical sciences during the late 1960s and early 1970s have spawned a major technological revolution popularly referred to as biotechnology. Unlike many of the technological innovations introduced during the American Industrial Revolution, which was not a revolution in science,[3] but rather one that was built on practical invention, the biotechnological revolution derives directly and immediately from the pure sciences. Once genes were capable of being excised, spliced, and transplanted into foreign cells, the power of the technique as an industrial tool was patently obvious to those who played a central role in its discovery. But the pioneers in molecular biology were not seeking an industrial revolution. Far from it, they stumbled on to one while in hot pursuit of new methods to study the genetics of higher organisms.

Several terms have been used to describe the changes in science and technology emanating from the biological research laboratories, including biotechnology, bioengineering, genetic technology, genetic engineering, recombinant DNA, and some more esoteric names like algeny and plasmid engineering. However, many of these terms do not arise *ex nihilo* but have an ancestry both in the processes they describe and in their etymology.

The term *biotechnic* appears as a fleeting thought in Lewis Mumford's extensive treatise on culture and machines titled *Technics and Civilization*.[4] In this work Mumford examines the development of technology through a millenium of modern history. The framework he presents for conceptualizing technological change is a useful guide for positioning biotechnology in the historical landscape of major industrial revolutions. Building on a typology derived from Patrick Geddes, Mumford posits three overlapping and interpenetrating periods of technological transition. Eotechnics represents the period defined by the handicraft industry, where human production was based on the use of natural materials available on the earth's surface, such as water and wood. Paleotechnics refers to the stage of mass production where subterranean materials, such as coal and iron, were appropriated. Finally, the third stage, which Mumford calls neotechnics, involves the development of electric energy, alloys, synthetics, mass communication, and the application of the concepts of science "to every phase of human experience and every manifestation of life."[5]

Each stage is associated with a period of human history and a family of technologies that share similar energy sources and correlate with a unique material basis and mode of production. By postulating these stages, Mumford can highlight the broad outlines of technological change in civilization.

Each phase . . . has its origin in certain definite regions and tends to employ certain special resources and raw materials. Each phase has its special forms of production. Finally, each phase brings into existence particular types of workers, trains them in particular ways, develops certain aptitudes and discourages others, and draws upon and further develops certain aspects of the social heritage.[6]

According to Mumford, the early stages of industrialization subjugated life processes to the machine. But during the neotechnic phase, the objects of mechanization began to be influenced by organic forms. This means that the animate world was beginning to serve as a model for technological design. The ear, the eye, wings, and vocal cords all became important symbols and even templates of mechanical invention. "Machines, which had assumed their own characteristic shapes in developing independent of organic forms, were now forced to recognize the superior economy of nature."[7]

Mumford not only brought attention to the metamorphosis of technics and its interrelationships with culture, but he anticipated human factors research. Instead of humans adapting to a fixed and ponderous technology, he saw instances in which the forms of biological life guided the purposes and structure of mechanical invention. For Mumford, this was an optimistic sign and one of the few counterforces to the rise of technological devices with dehumanizing capacities.

One of the principles of organic life is functionalism. Animals possess what they need to function and survive in some segment of nature. What lacks function, slowly disappears from the species. Writing in the 1930s, Mumford cites the trend in technology design of stripping the machine to its essential functions.

It is a first step toward that completer integration of the machine with human needs and desires which is the mark of the neotechnic phase, and will be even more the mark of the biotechnic period, already viable over the edge of the horizon.[8]

He introduces the term biotechnics to describe a hopeful trend whereby technology, "instead of benefiting by abstraction from life, will benefit even more greatly by its integration with it."[9] Mumford brought attention to the possibilities of technology for becoming life-affirming in its function, its aesthetic, and its physical coexistence with humans.

The term *biotechnics* is a hybrid concept derived from the terms "biology" and "technology." When Mumford introduced the term he was searching for a way to describe how living things can guide technological progress. Fifty years later, the world is a different place. Science and technology are now not so much guided by, as they are guiding, the research and invention of living forms. Life forms have become the substrate of a new technology driven by a mechanistic paradigm. Instead of living things standing in opposition to mechanical invention, as Mumford viewed them, they have become invention.

Edward Yoxen writes:

Our image of nature is coming more and more to emphasize human intervention through a process of design. More and more, genes, organisms, biochemical pathways and industrial bioreactors and processes can be realized according to a prior specification. Now, the essence of life is its constructability.[10]

Under the new scientific rubric of molecular genetics—currently the most fashionable framework for the analysis of living organisms—the gene has become the central concept both for the science and its industrial applications. The approach taken by traditional cell biologists has been overshadowed by genetic engineers. Some geneticists go so far as to view the cell as a convenient but not necessarily permanent home for genes that seek to perpetuate themselves.[11]

Within genes lies the essence of living forms, the *élan vital*, according to the current scientific fashion.

This perspective is supported by the following popular, albeit exaggerated, scientific claims. Genes direct the production of biological matter. They determine the physical characteristics of the life form. They are the basis of all change in organic life through the process of mutation. Increasingly, new genes are being discovered that are cited as a principal factor in the etiology of a disease. And finally, some scientists are pursuing sociobiological theories that posit a genetic basis for social and psychological behavior in humans.

An industrial revolution invariably influences the way we think about the world. Many new scientific concepts eventually filter into the popular vernacular. The changes in language occur through two processes. First, the instruments of technology arising from scientific discovery establish new relationships among people and between people and nature. These new relationships are soon reflected in the creation of symbols and images for comprehending and describing the social and physical elements in our environment. Second, the ideology of the industrial revolution is transmitted through the literature of the intelligentsia into the popular culture (novels, mass media, and lay renditions of science).

A new vocabulary is also needed to mark the differences between the older and newer forms of biotechnology. I shall use the term *gentechnics* (or *genetic technology*) to refer to those techniques that are used to rearrange bits of genetic information. I shall use the term *biotechnics* (or *biotechnology*) to refer to the class of procedures that operate on the entire organism, its genome, or any of its parts. Thus, in my terminology gentechnics is a subset of biotechnics. I am using biotechnics quite differently than Mumford, who defined it as the impact of biological form on technology.

Traditional biotechnics (old biotechnology) involves the external manipulation of organisms (temperature, acidity, nutrients) and therefore does not intervene in the fine structure of the internal controls. With the discovery of genetic engineering, biotechnics has been transformed. Gentechnics has introduced a new stage of biotechnics where internal and external controls of life forms can be imposed simultaneously. To pursue this point further, I shall discuss the role of the fundamental unit of life, the cell, as the emerging symbol and material basis for a new mode of production.

CELLULAR MECHANICS

The predominant metaphor for the cell in the age of modern biotechnology is that of a complex biochemical factory. Cells were appropriated for human use thousands of years before the industrial revolution (see Chapter 2). But not until the 1970s did scientists learn how to redesign the cell's architecture and appropriate its internal biochemical processes for production. The cell can be adapted to synthesize products that were not part of its telos or evolutionary plan. Cellular mechanics consists of those principles that define the cell's system of production.

The key discoveries that contribute to the mechanistic model of the cell span a period of nearly 40 years. In 1953 the genetic code was deciphered and the role of DNA (deoxyribonucleic acid) in the cell's synthesis of proteins was postulated and subsequently confirmed.

By 1960, messenger RNA (ribonucleic acid) was discovered. These molecules carry information from DNA to certain production sites of the cell called ribosomes where polypeptides are manufactured and fashioned into proteins. In 1967 an enzyme was isolated that causes DNA chains to join together (DNA ligase). Three years later another group of chemical enzymes were isolated called restriction enzymes (endonuclease). These enzymes cut DNA strands at specific sites.

Between 1972 and 1973 the newly discovered enzymes were used in the first experiments performed involving the introduction of foreign DNA into bacteria by cutting and splicing gene fragments. This class of experiments was named "recombinant DNA" expressing the fact that genetic information (DNA) is recombined in the test tube (*in vitro*).

The symbols, metaphors, and models that grew out of the recent discoveries in molecular biology treat the cell like a complex machine.[12] In part, this may be explained by the desire of scientists to communicate to an audience that lacks familiarity with the arcane language of biochemistry. But it also reflects the viewpoint within science that the cell has been brought under human control. The language of the machine is the language of power, determinism, causality, matter in motion, energy, and of controlling forces. The mechanistic materialism of modern genetics is quite distinct from the mechanistic views of physicists in the seventeenth and eighteenth centuries. The Newtonian world is characterized by material bodies in constant motion subject to external forces in a stochastic but deterministic system. The modern conception of the cell is based on a cybernetic materialism involving information transfer, energy, feedback signals, primary and secondary matter, synthesis of molecules, replication, and reproduction.

In the language of the new cybernetic materialism of molecular genetics, genes or DNA are referred to as "bits of information" or the carriers of the "code" for the production of proteins. These proteins form the "bricks" of the cell. Cells are the basic units of life; all microbes, plants, and animals are built from these tiny "chemical factories."[13] DNA is said to "unzip." DNA synthesizes messenger RNA (mRNA). The mRNA leaves the cell's nucleus and enters the ribosome, which begins the translation of its encoded information into chains of polypeptides that make up proteins. Ribosomes have been characterized as the cell's workrooms or its anvils where proteins are assembled. Their action is likened to that of a sewing machine; a more appropriate metaphor might be that of a laser reader. Genomes are spoken of as being mapped, DNA as being read or sequenced. There are classes of regulator genes with well-defined functions: promoters, operators, terminators, attenuators, cap sites, retro-regulators, translational control sites, switch sites, etc. A *Scientific American* article began with the statement: "A microorganism is a finely integrated machine that has evolved to serve its own purposes: survival and reproduction."[14] The term "gene machines" has been introduced to describe devices that synthesize specified stretches of DNA automatically. The frequently used expression "reprogramming microorganisms" exploits the language of computers to characterize human-induced changes in the DNA "code."

THE CELL AS A FACTORY

The symbol of the cell as a machine or factory unifies several key elements of the revolution in molecular genetics. It forms a bridge between the micro and macro aspects of biotechnology. The factory metaphor can be applied to the cell, the higher organism, the farm, and eventually to the urbanization of agricultural production.

Taken literally, the symbol describes the cell as a place of production. The cell's "machinery" can be appropriated to produce many useful substances, including pharmaceuticals, human proteins, and specialized chemicals, by incorporating into the cell the DNA template for the product desired. There are four requirements.

First, the foreign DNA must be incorporatable into the cell. Second, the cell must be capable of reproduction. Third, it must be capable of reproducing the DNA that has been implanted. And fourth, the inserted gene must be functional and capable of synthesizing a product.

We may draw an analogy to a factory that manufactures machine parts from metallic ore. Only one phase of the process, which involves the molds, determines whether the final product is a bolt or a drive shaft. Simply by changing the molds a new product may be fashioned from the raw materials. Similarly, the cellular machinery of the single-celled bacteria *Escherichia coli* (*E. coli*) can be used to synthesize human growth hormone or insulin once the gene for the appropriate protein and its regulating sequences are transplanted.

Anyone who has worked a factory assembly line knows the significance of speed-up. The line speed of the assembly process is increased thereby demanding higher worker productivity. A similar concept may be applied to bacterial production. Biotechnologists insert a gene into a bacterium that synthesizes a commercially useful product. For the process to be economically viable for large-scale production the rate of biosynthesis must be sufficiently high. Industrial scientists study how to amplify or enhance the cell's products by introducing multiple copies of a gene or by controlling environmental factors. As a "factory" the cell must be brought to peak efficiency.

Bacteria are reproducible and disposable "factories" of production. Once biosynthesis is completed in a large-scale fermentation system, the spent organisms are retired from production and relegated to waste. These bacterial production units usually have a life span covering the manufacture of one product cycle. As bacterial cultures grow so rapidly, enough is set aside for the next cycle of production.

Prior to the discovery of recombinant DNA, the cell's inventory of products was the outcome of its evolutionary development. Now, scientists can retrofit the cell's production apparatus with new instructions. Some bacterial species exchange DNA naturally and therefore acquire new chemical outputs. But the cell's genetic architecture limits the kind of DNA it could incorporate under natural conditions. A bacterium would gain no evolutionary or ecological advantage in incorporating human genes and synthesizing a human protein. The development of cellular mechanics or cell engineering is thus a critical phase in the emergence of commercial biotechnology.

LIFE FORMS AS PRODUCTS OF MANUFACTURE

The symbol of the cell as a "system of production" has legal implications. The awarding of patents to processes or objects of manufacture is a right granted by the Constitution. However, patenting is generally excluded for natural objects that do not bear the stamp of human intervention. One cannot simply culture a bacterium that synthesizes a useful enzyme and patent the product. Of course, there are always gray areas. If a unique or innovative process is used in the isolation of the organism, the courts have declared the process and the bacterial product patentable subject matter. The U.S. Patent and Trademark Office is responsible for deciding whether the organism's biological production process is natural and therefore unpatentable or whether sufficient human alteration of the cell has taken place to qualify it as an "object of manufacture." From the legal standpoint, the distinction is drawn between the cell as "natural producer" and the cell as "human-crafted" producer. The fact that bacteria, unlike machine parts, are self-reproducing entities has no bearing on the legal issues surrounding patentability (see Chapter 3).

ANIMAL PRODUCTION

The term "factory farming" connotes the restructuring of the traditional methods of animal husbandry used for millennia to breed farm animals for food consumption. In his book *Animal Liberation* Peter Singer summarizes the philosophy behind the new system of animal production: "Animals are treated like machines that convert low-priced fodder into high-priced flesh, and any innovation that results in a cheaper 'conversion-ratio' is liable to be adopted."[15]

The application of the industrial paradigm of manufacture to the farm has led to a highly mechanized system of raising, breeding, and slaughtering food animals. Far from being viewed as sentient beings with interests, a telos, or a species nature, food animals are simply "inputs" or "feedstock" in the production process. The morality of factory farming has been a cause célèbre for the international animal rights movement.

Two new phases of factory farming are introduced through genetic engineering. Both reinforce the view that the animal is part of the "machinery of production.'' The first phase involves redesigning the physical structure of the animal by direct manipulation of its fertilized egg. Genetic instructions determine how many teats a cow has, or the size of a chicken's breast, or the surface to volume ratio of a pig. These morphological characteristics are now subject to radical modification. Fertilized eggs have been artificially altered with foreign DNA to create new hybrid animals with desired phenotypes. The boundary conditions of natural procreation have been eliminated. The result has been a burgeoning industry of animal production through genetic reprogramming. This novel form of animal production is called transgenics, or the crossing of genes from different species to create chimeras.

The animal itself can also be viewed as a system of production because: (1) it reproduces itself; and (2) it produces valuable food for human consumption by changing raw materials (feedstock) into complex proteins. The factory symbolism carries over to the multicellular organism. By administering hormones, steroids, and other veterinary drugs, two forms of control are achieved; the control over the animal's own protein products and control over its reproductive cycle. Research is underway that examines how to increase the fertility of animals during a lifetime of procreation. Other research is centered on increasing the amount of milk produced by a single cow or increasing its ratio of lean to fatty meat.

The concept of the ''cellular mechanism'' has also been a useful heuristic for the research community. Edward Yoxen recalls a description of an organism from a course he took on cell biology. ''Organisms are self-assembling, self-maintaining, self-reproducing machines that operate at room temperature and pressure.''[16] Yoxen views the paradigm of the cellular machine as vital to the organization of research in molecular genetics. But his appreciation of the value of a mechanistic reductionism of life is met with some remorse.

Organisms are merely systems, and can be studied as systems, reducible in the last instance to a particular kind of logic. The study of process and form has been given up for the analysis of structure and linear order. It is the view of living things far removed from everyday experience and the aesthetic appreciation of shape, elegance, or anatomical subtlety, and equally far from naturalists' understanding of lifestyle and habitat. It is a biology built on fundamental abstractions such as the notion of a universal code, the idea of information, the metaphor of a programme controlling cellular activity.[17]

This form of biotechnics transforms our image of life into mechanical order. Richard Dawkins refers to all living forms as ''survival machines.''[18] The popular tinker-toy model of the DNA molecule, the twisted ladder (sugar-phosphate backbones) with interconnecting rungs (base pairs), helps to fix the popular image of the cell as a complex mechanical system with interchangeable parts. With the discovery of recombinant DNA technology, scientists speak about the fungibility of genes. In theory at least, any gene can be repositioned in a new organism. Interchangeable machine parts, the key to the development of modern industrial society, has its counterpart in the age of gentechnics where DNA can be moved to and from cells.

Contrast this with Lewis Mumford's vision when he postulated the role of biological processes in advancing the neotechnic age.

In the biotechnic order the biological and social arts become dominant: agriculture, medicine, and education take precedence over engineering. Improvements, instead of depending solely upon mechanical manipulations of matter and energy will rest upon a more organic utilization of the entire environment, in response to the needs of organisms and groups considered in their multifold relations: physical, biological, social, economic, esthetic, psychological.[19]

The final aspect of the new cellular mechanics I shall address is the role biotechnics is playing in the geographical transition of agricultural production from rural spaces to urban centers.

DECONSTRUCTING THE FARM

A revolution in scientific understanding of the cell is having a profound effect on macro-phenomena, such as food production. We have always acknowledged a distinction between agricultural and industrial modes of production. In the former case, soil, water, sun, and germ plasm represent the essential factors. The preeminent symbols of industrial production are the smokestack, assembly line, and fossil fuel energy sources.

Farming has been substantially mechanized in the twentieth century, specifically in soil preparation, irrigation, harvesting of crops, and animal husbandry. However, the basic factors of production have not changed, although additions have been made in the form of chemical pesticides and fertilizers, biological pest control agents, and new germ plasm. But food by and large is a product of soil, sun, and rainfall, along with the process of photosynthesis. So long as these factors are required, the farm will not be totally outmoded, although it may look very different from the small, labor-intensive family farm rooted in the Jeffersonian tradition.

Some analysts of the biotechnology revolution, including Yoxen, Kloppenburg, Kenney, and Doyle,[20] use the term agricultural restructuring to describe changes resulting from genetic engineering of plants and animals. The examples offered suggest that agriculture will face even higher forms of mechanization, more chemically intensive farms, and fewer but larger tracts operated by multinational corporations. In its 1986 study on the changing structure of American agriculture, the Office of Technology Assessment (OTA) identified six factors that represent structural alterations in the agricultural sectors: (1) number and size of farms; (2) contractual arrangements; (3) control of management decisions; (4) off-farm employment; (5) extent of tenancy; (6) ownership of farmland.[21] These factors identify an increasing trend toward the corporatization of the farm and the elimination of farming as a "way of life." Despite these changes the concept of the farm as a rural land mass that turns germ plasm into food through the use of soil, sun, and rainfall persists.

But there is another transformation that is taking place, albeit slowly, that will so radically alter agriculture that the primary components of farming (soil, sunlight, and rainfall) will be replaced. Under these circumstances, rural areas will not be required for food production. When this occurs, agriculture will be fully realized as a factory system.

For this transformation to take place, the concepts of food and food production will have to be substantially changed. Food will be viewed as the output in a production process that changes a feedstock into proteins, vitamins, and carbohydrates, and creates a product that people or animals either require for growth and nourishment or desire or would consume. There are many ways that these conditions can be met without farming the land. Jack Doyle describes one of these methods in his book *Altered Harvest*.

In the early 1970s, several of the world's major oil and chemical companies, including Exxon, DuPont, British Petroleum, and Standard Oil of Indiana, were at work at what they thought would be a new cheap source of animal and human protein . . . a substance called "single-cell protein." This protein is made by microorganisms and extracted from their dried cells. The microbes used in this process—fungi, bacteria, or algae—are given a food source (also called a feedstock) and are "grown up" in large batches in fermentation tanks. The multiplying microbes then serve as tiny "protein factories" as they reproduce, and the protein extracted from them can be fabricated into livestock feeds, used as food additives, or even turned into products that look like conventional foods.[22]

The fermenter, bacteria, engineered genes, and feedstock would replace soil, sun, rainfall, and germ plasm as the primary factors in production. The application of biotechnology to food manufacture overcomes the last obstacle for the rationalization of rural existence. The factory farm can then be appreciated not just as a metaphor, but as genuine realization of the transforming possibilities of the new bacterial modes of productions.

A second example of agricultural restructuring that may result in a rural to urban demographic transition is the development of tissue culture in conjunction with hydroponics. Plants are being produced from single cells or tissue extracts in culture in order to provide genetically identical plants with special characteristics such as an ability to survive high salinity, frost, or high-acidity soil. However, once the roots and shoots of these plants are developed from protoplasts, the use of irrigation channels nourished with the appropriate chemicals and artificial light may vitiate the need for a land-based food production system. On the basis of this potential, a giant food conglomerate like Campbell's Foods can justify the claim it makes to its stockholders: Biotechnology will enable us to design food to almost any specifications we want.

GENES, SYMBOLS, AND CULTURAL VALUES

Each scientific and technological revolution brings with it changes in social and cultural interpretation. It provides a new lens through which science and eventually the popular culture views reality. Philosophy, literature, social science, and art are strongly influenced by changes in scientific theory and explanation. Science carries in its wake a metaphysics that eventually is adopted into language, reason, and discourse. This can be amply illustrated in the Copernican, Newtonian, Darwinian, Freudian, and Einsteinian revolutions.

Since the structure of DNA was revealed in 1953, a social iconography of life has been emerging concurrently with the scientific *Weltanschauung*. In the scientific vernacular, the gene is a certain length of DNA, the smallest stretch of nucleotide sequences that is capable, within the environs of the cell, of synthesizing a protein. It is generally recognized that genes do not behave like isolated monads. Their function depends on the protein structure within the environment in which they reside. There is a constant interplay between the gene and the surrounding field.[23] Nevertheless, for the molecular geneticist, the gene can be isolated, deciphered, cut and spliced, amplified, transplanted, and synthesized. It can represent nonsense or it can be a vital element in the transmission of hereditary characteristics.

In large part because of refinements in the methods of identifying and sequencing genes and because of the significance of genetics in the transmission of biological traits, the gene has secured a preferred ontological status in biology. Like other concepts—such as the atom, the ego, relativistic length, and species—in their respective theories, the gene is a primary unit in the conceptual architecture of molecular genetics.

But in a broader cultural context the gene serves as a bridging symbol between science and lay society. Certain concepts in science do double duty. They become part of our cultural vocabulary while they also serve the enterprise of science. In the popular understanding of biological nature, humans are faced with two determinations. Genes represent the deep structure of inner life—that which defines our fixed nature. The environment signifies all that is not fixed from within.

Since the concept of the gene was introduced into popular terminology, it has been met with mystery, awe, and even veneration. The existence of our genes predates our physical being. Our genetic composition is the essential claim on us by our progenitors and represents the continuity of our ancestral line.

In common vernacular, the phrase "it's in his genes" signifies that some human quality or attribute is beyond our control. It is part of our "fixed" nature. Bemoaning our genes is akin to regretting our fate. As more and more aspects of human disease and behavior are explained by reference to genetic determinants, the inevitable outcome is that people ascribe a greater role to providence in their life struggles.

I would venture a guess that as we place more emphasis on genetic causation, people will be less willing to accept personal responsibility or external factors as explanations for their condition. If we believe that there are "criminal genes" then the concept of "guilt" for violent crimes may be as peculiar to us as attributing personal responsibility for blue eyes or tall stature.

The complex interaction between genes and the environment at the cellular and organismic levels are easily trivialized by popular presentations in favor of a two-value universe—nature and nurture. This popular conception of a bimodal reality serves a function by simplifying the elements of explanation, at the expense, however, of negating the importance of interactionism.

The human genome is the subject of considerable attention. Efforts are underway to sequence most of the estimated 100,000 genes. A map of the human genome is to geneticists what a map of the celestial sphere is for astronomers or a map of elementary particles and their energy states is for physicists. For the staunch advocates of the human genome initiative, the inevitable outcome of the genetics research program is the determination of the structure, function, and identity of all genes in an organism, but each studied separately. Science, of course, forever regenerates. Another research program will emerge that will be far more synthetic and less reductionist in its approach and will begin to view genes in complex environments as part of dynamic interactive systems.[24]

Our society will greet knowledge of the human genome with ambivalence. What will it mean to know our genes? In some respects it will be to know our fate. Are we predisposed to alcoholism or schizophrenia? Why are we short? How much of our height can be attributed to diet? We already attribute a significant component of height to genetics. I am quite satisfied to know that my height limitations are derived from my parents. If I have the genes of a short or medium-statured person, what significance will there be in having precise knowledge of my genetic makeup? The one difference it could make is if the knowledge we possess will result in opportunities to overcome our genetic constraints. The ability to modify human growth is a good example of the dilemmas that may result from mapping the human genome (see Chapter 11).

Genes are rapidly gaining a unique and arguably undeserved status, among scientists and eventually throughout society, as the quintessential factor of human existence. Many human characteristics that used to be attributed to life's lottery, the environment, or developmental factors are now being assigned a genetic cause. Some of these changes will be received favorably, particularly when a dread disease is at stake and intervention for the prolongation of life is possible. We are no longer fated to die because we lack a gene to produce enough insulin. First animal insulin and now human insulin produced through biotechnology have given diabetics who are otherwise untreatable through diet a reasonably normal life.

However, as science reveals more about the role human genes play in developmental biology and the etiology of disease, there will be new choices for individuals and society that will challenge our concept of personhood, our idea of privacy, our sense of rightful intervention into the human lifeline, and our belief about what we must accept as fate. New technological possibilities in human genetics create new expectations, diagnoses, treatments, risks, and opportunities for human exploitation.

Genetic technology, for example, is forcing us to reexamine the legal definition of personal identity. Gene sequencing has been introduced into the courts as a failsafe method of identifying a crime suspect in cases where human cells have been left at the scene of the crime. It has also been used to identify children who have been stolen from their biological parents and corpses that have been severely decomposed. If this approach, often misrepresented by the term "genetic fingerprinting," becomes an acceptable practice, the traditional view of personal identity, based exclusively on phenotype, will be challenged.

The concept of a fixed, unalterable human biological essence for each individual—an essence transmitted through the bloodline—is being deconstructed by virtue of modern genetics. Like the classical views of absolute space and time, which eventually displaced Aristotelian views of telos and becoming, and were themselves transcended by the Einsteinian revolution, the social knowledge of human heredity will also undergo a transformation. The gene has become a potent cultural symbol—one that embodies hope and despair. Hope comes from the possibilities for new medical treatments for inherited disorders, while despair sets in when new knowledge magnifies our vulnerabilities that we have no power to change. Such opposing tendencies are represented more significantly in the genetics revolution than in any other technological revolution over the past millennium, with the possible exception of nuclear physics. In the following chapters I examine how the revolution in applied genetics (gentechnics) has been introduced to society. During the early stages of bioindustrialization we find periods of regulatory malaise, public confusion, and anxiety over whether to welcome or curse genetic engineering. By the early 1980s a great campaign had been started by major corporations, industry trade associations, state governments, and universities to promote the biotechnological revolution first to the investment community and then to the American people, promising a cornucopia of improvements to civilization. Subsequent chapters in this work will assess some of these expectations and the public's response to them.

SOCIAL GOVERNANCE OF TECHNOLOGY

Throughout the many important technological revolutions that have occurred in the industrial world, it is difficult to identify a single example that illustrates a social commitment to explore the dark side of technology before casualties appear. There is an important reason why this has been the case. The eagerness to exploit a technology's benefits greatly exceeds the social commitment to assess its safeguards. At the root of the imbalance between promotion and control during the early stages of technological innovation lies the economics of development. The incentive system, particularly but not exclusively in capitalism, disproportionately favors bringing the technology into use first, thereby creating markets, wealth, labor-saving devices, and weapons of war, and evaluating its social and ecological impacts later.

In most instances, those who support the social management or regulation of a technology must demonstrate why that is warranted. Generally, a new technology is innocent until proven guilty. It is deemed progressive unless proven regressive. This has been the guiding caveat of technological change.

Since the 1970s, when Congress enacted a significant body of environmental legislation, there has been a modest but not insignificant shift in the burden of proof with respect to the introduction of new products, processes, and technologies into commerce and industry. The concept of prospective technology assessment was introduced into some decision processes. For example, in the National Environmental Policy Act (1970) there is a requirement that major federal projects undertake an environmental impact assessment subject to public comment. The mere requirement of an impact assessment imposes no obligation to modify the outcome of the project. To the contrary, it is only a recognition that consideration of adverse impacts is a legitimate concern in setting public policy. Nevertheless, the requirement nourished public advocacy and democratic participation.

The growth of risk assessment is another indicator that anticipatory planning of adverse technological impacts is gaining favor. In 1972 Congress established the Office of Technology Assessment to "help legislative policy makers anticipate and plan for the consequences of technological changes and to examine the many ways, expected and unexpected, in which technology affects people's lives." The political activism of the 1960s stimulated many legislative and institutional reforms over the next decade that give greater weight to the health and environmental consequences of technological change.

Despite the idealism behind the progressive legislation and the positive steps that were taken, there continues to be a vast imbalance between the human resources allocated to technological innovation and the social resources allocated to the study of technological impacts. Moreover, the shift to a more conservative philosophy of American government has brought with it a strong reaction to new regulatory initiatives.

By exploiting the theme of overregulation in favor of laissez-faire entrepreneurship, the Reagan administration stunted the progress of the health and safety initiatives and critical technology assessment programs that marked the 1970s as the golden age of environmental reform. While the antiregulatory sentiment gained converts among intellectuals, a new technology was born. The unlikely marriage between genetics and industrial process has become the latest symbol of human progress. Biotechnology finds itself among several elite high-tech industries, including microelectronics and robotics. But for those concerned about the hidden liabilities of new technologies this is a unique opportunity. Industrialized nations have had decades to reflect upon the adverse consequences of synthetic organic chemicals and nuclear radiation. We have watched the internal combustion engine and fossil fuel power plants transform our precious natural resources into generators of illness and ecological damage. The mistakes of our technological history are apparent before our eyes. They serve as a warning and guidepost for present and future human activities.

BUILDING A NEW TECHNOLOGICAL PERSPECTIVE

Each technological revolution is shaped by a unique historical context. To understand the possibilities for human intervention in guiding the course of an industrial revolution we must give considerable attention to the social and economic conditions that exist during its early stages of development. The emergence of atomic energy came at a time when science was hostage to the threat of fascism. The mere thought of a world under the swastika delayed for 20 years a serious assessment of the health and environmental effects of the nuclear fuel cycle. Wartime imperatives legitimized governmental coverups of radiation hazards in atomic testing programs. Recent investigations of U.S. nuclear fuel processing facilities have disclosed a pattern of deceit to the public and a wanton disregard for the environment, all under the mantle of Cold War politics. In contrast, biotechnology has been born in peacetime. It is the first major technological revolution in the aftermath of the Vietnam War. Unlike its predecessor, atomic energy, which like a convex lens focused the nation's industrial and intellectual resources toward a single goal,[25] genetic technology is viewed as a major contribution to a broad-based industrial revitalization of the United States. It is widely divined that biotechnology will create new world markets for American industry that will make up for the foreign penetration into its domestic markets.

Although there are military interests in biotechnology, and they are growing, science and society are not being held hostage by an ideology that implores us to develop first in the name of national security and ask questions later.[26]

There are also important differences between the early development of synthetic organic chemicals and the creation of genetically engineered life-forms. In the latter case, scientists issued a warning that the work they had begun could yield hazardous consequences. This is in stark distinction to the early developments of industrial organic chemistry when no warnings were issued by the scientific community about the potential injurious effects of thousands of synthesized organic compounds.

In the early 1950s the moral responsibility of science was synonymous with its self-realization. Science was viewed both as inherently virtuous and as the engine of human progress. By the 1970s many scientists were influenced by the moral indignations of the Vietnam War and the arms race. The moral responsibility of science became a prominent theme that had evolved from the scientists' role in developing the atomic bomb and was applied to human experiments, war research, and subsequently to genetic technology.

The revolution in industrial genetics arrives at a time when public and scientific understanding of the impacts of technology has matured considerably. It is difficult to plead naivete given our historical vantage point. Nor is it reasonable to assume a priori that the industrial uses of biology will be significantly safer or more humane than the industrial uses of organic chemistry or nuclear physics. On intuitive grounds alone it seems quite implausible. Why should the manipulation of the basic chemistry of life, the gene, come at no social cost, when analogous transformations of the basic units of inert matter have created profound problems? To raise this question is not to make a judgment against a technological revolution. It is merely a way of punctuating the fact that technology comes in two denominations: assets and liabilities. Those sectors of the economy involved in technological development usually focus exclusively on the assets. The liabilities must be addressed by other sectors.

To guide biotechnology safely through a future path of potential liabilities represents an important moral responsibility for the public sector. This responsibility is predicated on the premise that options exist. For the individual, moral culpability for behavior presupposes free will. For society, moral responsibility toward technology implies that technological determinism is false. Another way of stating this is that technologies are not autonomous; they do not have a life of their own. While the pros and cons of this debate are argued interminably in academic circles, one's attitude about technology is grounded on principles of faith more so than logic or reason.[27] It is my belief that industrial America, through its experience with the atomic age, hazardous chemicals, ecological degradation, and occupational hazards, can draw upon the collective wisdom of largely well-intentioned people and guide biotechnology through a path safer than that of previous technological revolutions of a similar transforming character.

But good intentions are not sufficient in themselves. We need to know what to assess, what to control, and what to promote. We need to develop the appropriate institutional mechanisms for such assessment procedures. These mechanisms should be built on democratic processes that foster public involvement. At the present time, the scientific basis for assessing some of the most significant and recognizable outcomes of biotechnology simply does not exist. As an example, consider the intentional release of genetically modified organisms into the environment (see Chapters 6 through 8). Under what conditions would such a release be safe or hazardous? The problem has barely been articulated within appropriate scientific disciplines. It is not clear whether problematic cases can be anticipated if social and scientific resources are allocated to such problems. Something is demonstrably unsafe when the hazards are revealed. But in many areas of biotechnology, little is known about how to ascertain the social and ecological impacts at the anticipatory phase, prior to marketing products.

Under these circumstances we are faced with a dilemma. It is not known whether, or to what extent, prospective technology assessment can guide us along a safer path. A faith in reason would dictate that we try. Then we are confronted with the problem of resource allocation. What level of public and private resources should be directed at the task of detecting the possible adverse impacts of products or processes prior to entry into commerce? How do we translate good intentions into a sensible program of social guidance? How do we put into practice the idealism of "safer technologies this time"? I shall return to this theme in the final chapters of this book.

The first decade of the revolution in biotechnology provides a bellwether of its future direction. This book begins the task of chronicling the nascent stages of its development. We do not yet know whether the revolution in applied genetics will establish higher standards for civilization as a whole, will respect diverse forms of life and habitats, will liberate us from disease or enslave us to a genetic determinism, whether its achievements will be shared equitably, or whether its significance will be mixed with a favorable outcome to narrow interest groups.

The book is organized around three themes. Part I explores the industrial context by examining the origins of the biotechnology industry, the role patenting has played in stimulating commercial development and introducing new legal quandaries, and the commercialization of academic biology. Part II examines one of the most controversial aspects of biotechnology, the release of genetically modified organisms into the environment. It covers the evolution of federal policy from laboratory experiments to field trials of genetically modified microorganisms and plants and describes the social processes in place to review the first field experiment. Part III looks more generally at social controls and addresses the problems that arise from the fragmentation of regulation and the lack of consensus over the goals of biotechnology.

Part I

The Industrial Context

The Emergence of the New Biotechnology Industry

For the accomplishments of the still infant biotechnology industry, its promises of new procedures and new products, to say nothing of its ever more efficient production of goods, are awesome.

Daniel E. Koshland, 1985[1]

The biotechnology industry has repeated, step by step, what has happened in the petrochemical industry. If it is allowed to go further, like the petrochemical industry, it will become invulnerable to control. . . . Now, at this early stage (if, in fact, it is not already too late), we need to control what the biotechnology industry produces.

Barry Commoner, 1987[2]

This chapter examines the early development of the emergent industry that owes its existence to the new scientific advances in genetics. With ancestral roots in industrial microbiology that date back centuries, the new biotechnology industry grew rapidly but in discernible stages beginning in the mid-1970s. Each new scientific advance became a media event designed to capture investment confidence and public support. Market expectations and social benefits of new products were frequently overstated. It was part of the "geneticization" of the social mind. People were being prepared to see genetics as the next great advance in technological progress.

The new biotechnology industry benefited significantly from strong federal support, many state economic development initiatives, and aggressive university participation in technology transfer. By the end of the first decade the industry was faced with fewer than expected mass market products, public skepticism over certain areas of development, and conflicting regulatory issues. Nevertheless, despite unresolved societal questions, the nascent industry was beginning to fully integrate into established industrial sectors and show signs of healthier patterns of market growth.

BIOTECHNOLOGY: PROCESS OR INDUSTRY

The term *biotechnology* has been defined as "any technique that uses living organisms (or parts of organisms) to make or modify products, to improve plants or animals, or to develop microorganisms for specific uses."[3] The nomenclature came into use in the early 1970s when a new assortment of genetic and cellular techniques such as recombinant DNA, cell fusion, and nucleotide synthesis were transferred from academic to commercial laboratories. Plant selection, animal eugenics, antibiotics, and the uses of microorganisms in food preparation are some examples of traditional forms of biotechnology. Generally, they involve the use of whole organisms in production.

The modern generation of bioprocesses, which I have referred to as gentechnics, differentiate "new" biotechnology from "old" or "classical" biotechnology in that the former involves the transfer of specific fragments of genetic information from one cell to another.

Biotechnology is also used to describe an emergent industry. Paradoxically, there is no product or class of products that can be identified with this nascent industry. Because of this some analysts consider the term misleading.[4]

A 1984 study by the congressional Office of Technology Assessment stated: "It is important to recognize that there is no 'biotechnology industry.' Biotechnology is a set of technologies that can potentially benefit or be applied to several industries."[5] In a follow-up study OTA referred to biotechnology as a set of enabling technologies applicable to a wide range of industries.[6]

In actuality, a number of industries view biotechnology as one or more generic techniques involving cellular, subcellular, or multicellular units that can contribute to a more efficient means of manufacturing traditional products or can deliver new products to the consumer. Thus, while the chemical industry is organized around the manufacture of chemicals, biotechnology is associated with the production of microbial agents, proteins synthesized by those agents, plants, and animals, as well as pharmaceuticals and industrial chemicals. Biotechnology is also identified with medical therapies, such as human somatic cell therapy, and reproductive technologies, for example, germ line gene therapy. Table 2.1 shows a breakdown of 219 firms in seven industrial sectors pursuing applications in biotechnology around 1984.

The revolution in applied genetics has resulted in structural linkages between disparate industries such as energy and pharmaceuticals. These linkages are created through a new generation of R&D firms that are organized around the application of genetic techniques rather than product lines. The industrial affinities developed around genetic technologies are exemplified by the creation of new trade associations such as the Association of Biotechnology Companies and the Industrial Biotechnology Association, new trade publications such as *Genetic*

Table 2.1
U.S. Firms Pursuing Applications of Biotechnology by Industrial Sector, circa 1984

Sector	No. of Firms*	(%)
Pharmaceuticals	136	62
Animal Agriculture	62	28
Plant Agriculture	53	24
Specialty Chemicals & Food	47	21
Commodity Chemicals & Energy	33	15
Environment	25	11
Electronics	6	3

Source: Adapted from Office of Technology Assessment, *Commercial Biotechnology*
(Washington, DC: U.S. Government Printing Office, 1984), 71.

Engineering News and McGraw Hill's *Biotechnology Newswatch*, and new journals like *Trends in Biotechnology*. Sectors already involved in traditional biotechnology (e.g., fermentation and pharmaceutical industries) were revitalized and/or reorganized in response to the new developments in molecular genetics. This is a case where a scientific and technological revolution has helped to recast industrial boundaries, fostering entirely new pathways through which economic sectors interact.

REBIRTH OF A VENERABLE TECHNOLOGY

Broadly defined, biotechnology has been a part of the human repertoire of techniques since the dawn of civilization. The application of microorganisms to food preparation through a process of fermentation represents one of the earliest known uses of biotechnics. Fermentation was in use at least 8,000 years before the role of microbes in the process was discovered in the seventeenth century.[7]

The earliest evidence of beer-making has been traced to ancient Sumerians and Babylonians, possibly dating as far back as 7,000–5,000 B.C.[8] By 4,000 B.C. the Egyptians were using yeast to make leavened bread.[9] Ancient breadmaking and brewing of beer is depicted on the walls of a 5th-dynasty Egyptian tomb dating from around 2,400 B.C.[10] Fermented milk was known to the Sumerians, who most likely learned the process by accident. Milk stored in earthenware was inoculated naturally by bacteria present in the crevices of the container.[11]

Vinegar and dilute solutions of acetic acid made by a fermentation process mentioned in the Old Testament were used by Hippocrates as a medicinal agent.[12] Many fermentation processes applied to food preparation, such as the preservation of vegetables, predate recorded history.

The preparation of soy sauce, which involves the complex biochemical reactions of molds, yeasts, and bacteria, has been practiced in Japan for at least 1,000 years.[13] The fermentation of soybeans may have originated in Japan as a result of the introduction of Buddhism from China in 552 A.D. and the changeover to a vegetarian diet.[14] In the fourteenth century, people in various parts of the world were fermenting diverse grains to manufacture alcoholic beverages.[15]

The use of microorganisms for leaching metals from low-grade ores is also traced back to ancient societies.[16] Antedating the birth of "new" biotechnology, copper and uranium have been commercially produced by the leaching action of bacteria, predominantly of the genus *Thiobacillus*.

Although microorganisms have been used domestically and commercially for thousands of years, their role in these diverse processes was discovered barely a hundred years ago. With the development of the microscope and the discovery of bacteria, scientists rapidly learned about microbial action on foods, including causes of contamination.

The field of applied or industrial microbiology took off after World War I. Microorganisms were used industrially to produce organic acids by the action of molds, yeasts, and bacteria. Japan, which has one of the most advanced biotechnology industries in the world, was producing ethanol by fermentation in 1935 and using it as a substitute for oil. Between 1935 and 1938, the Japanese exploited the ability of radioactive radium to create genetically mutated organisms that were adapted for commercial use. They were pioneers in the application of mutant strains for industrial fermentation processes.[17]

By 1914 microorganisms had been applied to the treatment of domestic sewage. The "activated-sludge" process was developed in that year. It involved the inoculation of wastes with selected mixtures of microorganisms that efficiently break down the organic matter. Today, facilities using this process are commonly referred to as secondary sewage treatment plants.

Microbiology was introduced into the pharmaceutical industry after World War II for the production of antibiotics. Currently, this industry represents a substantial segment of the U.S. economy, producing, in addition to antibiotics, vitamins and human proteins. The sales of antibiotics alone represents a multibillion dollar industry.

Until the early 1970s, the biotechnology industry was dependent principally upon the use of natural organisms. There were limits on the extent to which microbes could be modified to improve production. Molds and bacteria had not evolved to adapt to commercial fermentation vats. Organisms could be altered by exposing them to chemicals or radioactivity, but the results were always haphazard and limited. Not until the discovery of recombinant DNA technology was it feasible to tailor-make microorganisms for specialized industrial tasks. The possibilities for redesigning microbial architecture are still in their infancy as gene splicing begins to reveal the principles behind the cell's internal program of protein synthesis.

From its origins in the 1920s, industrial biotechnology was constrained by nature's own boundary conditions. The industry was dependent on the discovery of organisms that were encoded with the genes of useful products. Sheep cells produce sheep insulin; human cells product human insulin. The exception was plant breeding, where grafting methods opened up a lucrative commercial field of hybridization decades before the new field of biotechnology arose.

The distinction between pre- and post-1970s biotechnology is analagous to the demarcation between analytic and synthetic chemistry. There was a time when industrial chemistry meant discovering uses for chemicals or their mixtures that are found in nature. With the discovery of chemical synthesis, it became possible to construct compounds that were unique to the planet and possibly the universe. Just as the discovery of synthetic chemistry revolutionized the chemical industry, according to most observers synthetic biology is having a similar effect on biotechnology. These projections are coming from people positioned on all sides of the political debates. Edward Yoxen argues that with biotechnology "the prospect of massive industrial change is imminent."[18] Marc Lappé refers to recombinant DNA as the "millennium in biology just as the discovery of nuclear fission was the millennium in physics."[19] The power to do good or evil in biotechnology, as in nuclear physics, according to Lappé, is "virtually limitless." The extraordinary power attributed to the new biotechnology is found in the interchangeability of genetic material across species. That is also the source of public anxiety.

COMMERCIAL EXPECTATIONS

The industrial organizations that were spawned around biotechnology developed financing on the promise of new products and new markets. The Office of Technology Assessment encapsulated the expectations of the fledgling industry in its 1984 report:

Biotechnology has the technical breadth and depth to change the industrial community of the 21st century because of its potential to produce substantially unlimited quantities of:

• products never before available,
• products that are currently in short supply,
• products that cost substantially less than products made by existing methods of production,
• products that are safer than those that are now available, and
• products made with new materials that may be more plentiful and less expensive than those now used.

By virtue of its wide-reaching potential applications, biotechnology lies close to the center of many of the world's major problems—malnutrition, disease, energy availability and cost, and pollution.[20]

It is particularly noteworthy that the OTA presented this untested industrial revolution as a salvo against problems of a global significance that thus far had evaded both political and technological solutions.

To the biotech companies and a financial community eager to invest in a virgin industry, the OTA's claims were important in counteracting negative public perceptions. Venture capitalists were thinking about biotechnology in terms even grander than IBM or Xerox. Early market studies were priced in excess of $50,000 despite the rapid changes taking place in the sciences and the unproven commercial successes of the technology.

To lure public investments, the biotechnology industry focused attention on several high-visibility products, including interferon, human insulin, clotting factor, and growth hormone. These were the products that drew significant media attention; they represented the "good-will ambassadors" of the new industry. The stakes were high for producing one of the so-called miracle drugs. It would have a ripple effect throughout the entire industry. Between 1976 and 1986 more than $3 billion was invested in the United States from public and private sources to bring these and other products to market. OTA estimated that U.S. industry invested between $1.5 and $2.0 billion in R&D for 1987, indicating the rapid rise of capitalization.[21]

From the late 1970s through the mid to late 1980s, the new biotechnology firms (NBFs) were not earning their incomes through product manufacture. These small biotechnology companies had licensing agreements with U.S. and foreign companies that generated operating income.

In anticipation of massive biotechnology markets, investment fever over biotechnology intensified during 1980. Concerns over regulation that plagued academic researchers in the 1970s had all but disappeared (or so it seemed) and confidence was spreading throughout the investment community in the application of recombinant DNA (rDNA) techniques. The media was quick to report carefully orchestrated news releases of the latest breakthrough in molecular genetics.

On January 16, 1980, the Cambridge, Massachusetts-based company Biogen held a press conference at which it announced its success in getting bacteria to produce interferon. Hardly a household word at the time, interferon had been isolated by scientists decades before and studied for its potential anti-viral activity. The microbial production of this natural protein was a boon to scientists who were eager to test its effectiveness on a variety of diseases from the common cold to cancer, and subsequently on Acquired Immunodeficiency Syndrome (AIDS).

Biogen's news release, which was skillfully orchestrated to arouse media attention, had a demonstrated impact on the company's stock value. The announcement of the scientific breakthrough was made before the results were published in the scientific literature. Sharp criticisms were directed at Biogen officials by members of the scientific community citing the unfortunate influence of business on the protocols of scientific communications. Arnold Relman, editor of the *New England Journal of Medicine*, known for his long-standing concerns over the commercialization of medical practice, published a critical editorial by Spyros Andreopoulos titled "Gene Cloning by Press Conference." The commentary emphasized the breaking of tradition and the erosion of professional responsibility.

> In place of published data, open to all for examination and critical review, we now get scientific information by press conference. The abrogation by scientists of the normal processes of scientific communication does not help science or the reporters covering it.[22]

The commercial prospects of interferon were touted by the media. The *New York Times* ran premier science journalist Harold Schmeck's story on prime first-column space under the bold headline "Natural Virus-Fighting Substance Is Reported Made by Gene Splicing."[23] By the summer, *Fortune* magazine described interferon as "the potential new wonder drug that is supposed to defeat viral diseases the way antibiotics conquered bacterial infections, and reported that Biogen's news release about the bacterial expression of interferon "helped make rDNA one of the hottest investment fields of the new decade."[24] Interferon's early tests on animals were expected to be replicated on humans once sufficient supplies of rDNA-interferon became available. The market for the new drug was estimated to be in the multibillion dollar range. The media publicity for a product of biotechnology so early in its testing stages was unprecedented. More than any single commercial application of rDNA, interferon captured the world's attention and promoted the wizardry of science, as exemplified by a statement in *World Press Review*: "The most spectacular potential in the [rDNA] field is human interferon, a highly praised anti-viral agent that could provide cures for diseases ranging from the common cold to cancer."[25]

By 1988, results of human trials of various interferons were reported in the scientific literature. These reports dampened any hope that science had discovered a miracle drug. There were studies that indicated interferon produced some anti-tumor activity for the treatment of leukemia, but also cited serious side effects.[26] Other studies indicated that nasal sprays with one type of interferon were not only an ineffective treatment for the common cold but exhibited toxic effects.[27] Interferon was tested on a wide variety of diseases, including hepatitis B, various lymphomas, rheumatoid arthritis, genital warts, multiple sclerosis, melanoma, breast cancer, osteogenic carcinoma, and AIDS. The development of rDNA-produced interferons has spawned a spectacular research enterprise, including an international society, federal and industrial research programs, and the *Journal of Interferon Research* started in 1981. At the center of this enterprise are a number of heavily invested companies with a "miracle drug" searching for a disease with an equally spectacular pharmaceutical market. In 1984 OTA listed 34 companies worldwide involved in interferon gene-cloning projects. Zsolt Harsanyi of Porton Industries wrote in the *Telegen Reporter* (September 1988) that "interferon represents a case of technology drive. Methods have been developed to produce large quantities of interferon at reasonable prices. Now it's time to find a use for the various interferons."

In his book *Gene Dreams*, Robert Teitelman quotes a pharmaceutical company executive on the issue of exaggerated claims.

There are very strong pressures from oncopolitics and oncoeconomics to make biotechnology more miraculous than it really was. By oncopolitics, I mean the need to keep the pump primed for cancer research; by oncoeconomics, the need to finance biotechnology companies.[28]

Only human rDNA-derived insulin could match the publicity afforded interferon during the early stages of the biotechnology revolution. Microbial insulin was a blue-chip investment. The international market was well-defined and dominated by a few firms. In 1981 the total U.S. insulin market was $170 million; the European market was $140 million. The OTA estimated that the world market would double in four years.[29] Diabetes was surely not on the decline worldwide and as a result there was substantial market potential for human insulin merely from the rise in population.

Anticipated agricultural applications of biotechnology also produced considerable investments in research. The financial stakes were every bit as high as in pharmaceuticals, and likely higher. In 1984 the OTA reported a U.S. seed market of $4.5 billion and a world market of $30 billion.[30] At the peak of the investment frenzy *Business Week* wrote that "a race is on to produce new varieties of crop plants that are resistant to certain herbicides. That would make it possible for a farmer to spray a field and kill anything except the crop plant."[31]

And so it went. In the late 1970s and early 1980s, newspapers and magazines helped construct an image of biotechnology as a revolution with enormous potential for improving the human condition. That image was cast out of results from the laboratory and based primarily on the word of scientist-managers, most of whom were new to the world of business. Cautious investors knew there was a great chasm between laboratory results and product development. The business and investment communities were receiving multiple levels of reinforcement. With such an extensive field of commercial applications for biotechnics, a few failures were not likely to affect the pace of development.

In 1980 not a single company made any earnings by manufacturing and selling a product developed through the new generation of bioprocess technologies. Genentech claimed earnings from research contracts supported by pharmaceutical companies. The NBFs were in a race for the first successful commercial products. They understood that investor confidence could not be sustained for more than a few years on promises, even if those promises had the imprimatur of eminent scientists. In the long run, products and markets were critical. For the purpose of creating a company image that satisfied the investment community, almost any profit-generating product would suffice. Table 2.2 shows a group of highly publicized first-generation products of biotechnology with estimated markets.

In the early stages of biotechnology, firms invested considerable sums into R&D to win the race for marketable products. R&D expenses greatly exceeded

Table 2.2
First-Generation Anticipated Biotechnology Products and Markets, 1978–1982

Products	Estimated Annual Market, 1984–1986
Foot-and-Mouth Disease Vaccine	$ >250 million (world)[a]
Human Insulin	$ >100 million (U.S.)[b]
	$ 630 million (world)[c]
Human Growth Hormone	$ 10–20 million (U.S.)[d]
Interferon	$ >2 billion (world)[e]
Animal Growth Hormones	$ 515 million (world)[f]
Thrombolytic Agents	$ 158 million (U.S. & Japan)[g]
Antihemophilic Factor	$ 360 million (world)[h]
Scours Vaccine	$ 100 million (world)[i]

[a] Office of Technology Assessment, *Commercial Biotechnology: An International Analysis* (Washington, D.C.: U.S. Government Printing Office, 1984), 163.
[b] Marjorie Sun, "Insulin Wars: New Advances May Throw Markets into Turbulence," *Science* 210 (12 December 1980): 1225–28.
[c] OTA, 1984, 121.
[d] Zsolt Harsanyi, *Investment Strategies in Biotechnology: The Race to Commercialization* (New York: EIC/Intelligence, 1984), 47.
[e] Gene Bylinsky, "DNA Can Build Companies Too," *Fortune* 101 (16 June 1980): 146; also, Sharon McAuliffe and Kathleen McAuliffe, *Life for Sale* (New York: Coward, McCann & Geoghegan, 1981), 72.
[f] OTA, 1984, 167.
[g] Ibid., 135.
[h] Ibid., 133.
[i] Harsanyi, 1984, 38.

Table 2.3
R&D as a Percentage of Operating Revenues for Leading New Biotechnology Firms, 1982

NBF	R&D as Percentage of Operating Revenues
Molecular Genetics	424
Monoclonal	423
Enzo Biochem	400
Genetic Systems	177
Hybritech	161
Genex	160
Cetus	143
Genentech	111
Biogen	72

Source: Office of Technology Assessment, *Commercial Biotechnology* (Washington, DC: U.S. Government Printing Office, 1984), 271.

operating revenues for many firms (see Table 2.3). Of 18 NBF-leaders in the industry, only three showed earnings between 1982 and 1983. Even these revenues from sales (including contract research) fell short of expenses. The investment analysts cited 1984 as a transition year during which the biotechnology industry was expected to generate $50 million from biotechnology-based products. Companies were supposed to begin turning a profit from product development, reducing their dependence on research contracts from large corporations.[32] In 1984, however, OTA reported:

Even the most mature NBFs at present have only a few products to generate revenues that can be used to cover operating expenses and provide capital for future growth. In order to generate revenue . . . NBFs in the United States are currently relying heavily on research contracts. The reliance of entrepreneurial firms on research contracts to generate revenue is almost without parallel, except perhaps for the small firms that do defense contracts.[33]

Four years later, it was still the case that no biotechnology company was able to report a profit solely from the sale of a product.[34] But the picture began to change in the late 1980s with products like Genentech's anti-clotting agent Tissue Plasminogen Activator (TPA), with 1989 sales of nearly $200 million, and human growth hormone (HGH), with 1989 sales of $123 million. Revised analyses from investment houses estimated a $2.5 billion bio-drug industry by the early 1990s.

GROWTH PATTERN OF THE BIOTECHNOLOGY INDUSTRY

New biotechnology firms began forming in the mid-1970s. Genentech is generally credited as being the first rDNA-based venture capital company. It was also the first NBF to issue public stock. However, firms like Collaborative Research and Cetus, both founded prior to the discovery of rDNA technology, began turning their attention to genetic engineering at the same time that Genentech and other firms were raising venture capital.

Currently, the composition of the biotechnology industry is diverse. Increasingly it has begun to look like a complex web. It consists predominantly of the NBFs, but also of large established companies (LECs) that have introduced R&D in applied genetics, major corporations with equity in small firms, partnerships, small firms with equity in one another, and nonprofit institutions such as universities with equity in private companies.

The increase in the number of new biotechnology enterprises (NBEs) between 1973 and 1987 was something of a phenomenon (see Figure 2.1). Financial backing of the new industry from venture capital and public stock offerings was accomplished with considerable ease, primarily on the basis of promises from revered members of the scientific community.

The primary techniques that spawned the new industry—rDNA and cell fu-

Figure 2.1
Aggregate Growth of New Biotechnology Enterprises (NBEs), 1973–1987

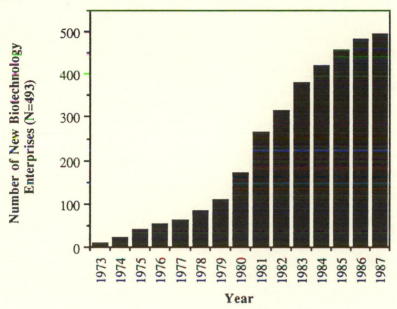

sion—were considered so revolutionary that it can be argued the industry was predestined to succeed. Early investments were in the techniques and the scientists who had control over them, not the products per se. Viewing the emergence of the industry from this point of view, it was irrelevant which products succeeded, failed, or met a social need. This was truly a technological revolution.

The failure of biotechnology was out of the question whether or not there were existing needs or favorable markets. The techniques would be used over and over again until successful products emerged. If the first generation of products proved unsuccessful or only marginally successful, other needs could be found to support a second generation of products. Unlike nuclear technology, where the major nonmilitary product is electricity, biotechnology is highly diversified. In 1988 nearly 40 percent of the R&D invested by biotechnology companies was in the areas of therapeutics and diagnostics.[35] There were dozens of potential products being considered covering a broad spectrum of uses in medicine.

The development of nuclear technology in the United States has been held to a standstill. Could the same conditions prevail in biotechnology? That is highly improbable unless the public perception of the entire industry was captured by a highly pejorative symbol. But within the broad parameters of the biotechnology industry some sectors have been threatened. They include the use of rDNA technology in producing genetically engineered microbes, herbicide-tolerant

Table 2.4
Biotechnology R&D Budgets in 1982 for Nine U.S. Firms

Firm	R&D Budget (millions)
Du Pont	120
Monsanto	62
Eli Lilly	60
Schering-Plough	60
Genentech	32
Cetus	26
Biogen	8.7
Genex	8.3
Hybritech	5

Source: Office of Technology Assessment, *Commercial Biotechnology* (Washington, DC: U.S. Printing Office, 1984), 74.

plants, and transgenic animals, and for use in human gene therapy. These are areas where the public has to be sold on biotechnology. In some cases where markets are not waiting to be tapped and the needs are not manifest, the bio-technology industry can create new needs. Moreover, a vigilant industry concerned about public perception leans toward obscuring the complex social impacts of any single product, since such an assessment could threaten the vitality of the entire industry.

In a 1984 study sponsored by the international investment house E. F. Hutton, it was estimated that research expenditures in biotechnology were $1 billion worldwide in 1980, $3 billion in 1984, and would grow to $6–7 billion annually in 1987.[36] Table 2.4 shows the magnitude of the investment capital infused into the industry through nine U.S. firms in 1982. The total R&D investment for these nine firms in that year was $382 million. A 1988 OTA study estimated the annual U.S. investment in biotechnology from federal and state governments and the private sector to be $4.3–4.9 billion, $1.5–2.0 billion of which was attributed to the private sector.[37]

The growth pattern of the NBFs and new divisions of long-established companies (LECs) that were formed to exploit biotechnology follows a bell-type curve (see Figure 2.2). The year 1973 marks the discovery of gene splicing. There is a slow growth of new firms until 1979 when the rate of formation rises exponentially, peaking in 1981 and tapering off rapidly from 1982 to 1987, with the exception of new growth in 1983.

A few established pharmaceutical and chemical companies, including Monsanto, DuPont, and Eli Lilly, have had biotechnology research efforts underway since 1978; however, most companies now commercializing biotechnology did not start until about 1981. A study by OTA reported that "the boom in bio-

Figure 2.2
New Biotechnology Enterprises by Founding Dates, 1973–1987

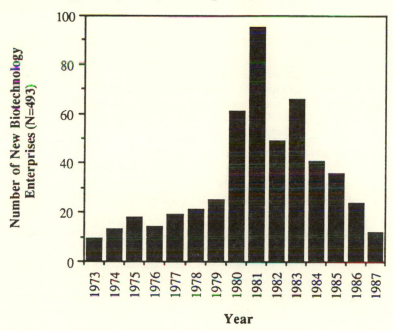

technology company formation occurred from 1980 to 1984. During those years, approximately 60 percent of current companies were created with nearly 70 new firms begun in 1981 alone."[38] The sharp decline in the formation of NBEs in 1982 is attributed to saturation in the field and an exhaustion of venture capital financing for startups.

INVESTMENT CAPITAL FOR NBFs

The principal sources of funding that financed the start-up companies in the biotechnology industry are: corporate capital (including equity purchases, contract research, and joint ventures), venture capital companies, public stock offerings, and Research and Development Limited Partnerships (RDLPs). The latter is a rather new form of financing arrangement that helped create a bridge between the investment and academic communities. I shall expand upon this briefly.

An RDLP is a financing mechanism designed to raise venture capital from the private sector to fund research and development projects. The entity consists of one general and one limited partner and allows a business to engage in research activities without paying for the activities out of retained earnings or borrowed capital. It allows firms to reduce their reliance on established companies and venture capital firms. The majority of the 300–400 RDLPs that existed in 1984 were formed after 1980. Between August 1982 and May 1983, the NBFs raised $200 million in RDLPs.[39]

Martin Kenney noted that:

The RDLP is an important mechanism for enabling startups to tap the capital of private individuals who are looking for returns not available from common stock. . . . Moreover, the startup acquires greater control over its fate—neither becoming a passive recipient of royalties nor risking the entire company on one go-it-alone product.[40]

RDLPs received strong backing from Congress and the Reagan administration. By October 1983 there were eight separate bills that exempted RDLPs from antitrust action and provided favorable tax status to them. They have been called the Marshall Plan for U.S. research and development.

A creation of the revised tax laws in the early 1980s, RDLPs were promoted as an alternative way to raise R&D capital for American campuses, improve private sector productivity, and enhance international competitiveness. Joseph W. Bartlett and James V. Siena cite special tax shelters and high investment income as the incentives for establishing a university RDLP. The RDLP offers the university a new and better approach to determine the commercial potential of its technology, fund the development of that technology, and bring it to the market. Because of the tax laws, the RDLP structure is an attractive vehicle for investors. It offers them the ability to deduct a large part of their investment and have future income taxed at capital gains rates. Finally, the RDLP offers the university the opportunity to minimize its entanglement with a commercial venture, yet at the same time increase its chances with a commercial venture, yet at the same time increase its chances of economic rewards.[41]

The Office of Productivity, Technology and Innovation (OPTI) was created in the Department of Commerce in 1981 by the Reagan administration. OPTI advocated the use of RDLPs at universities first as a means of generating nonfederal sources of research funds, thus reducing their dependence on federal dollars. Second, it was supposed to help accelerate the transfer of federally developed and funded technology for commercial use.[42]

Some observers noted quite correctly that once a university establishes an RDLP, there will be increasing pressure from investors for the funded research to be guided toward product development. As a limited partner in an RDLP, the university is a strong selling point to investors. It offers its name and reputation that may even be useful in marketing a product. From the university's vantage point, there is a new source of research funding and an instrument for capitalizing on the scientific discoveries of its faculty.

After a strong wave of venture capital, corporate, and RDLP financing for the biotechnology industry, NBFs began selling shares of the companies to the

Figure 2.3
States Leading in the Formation of NBEs, 1973–1987

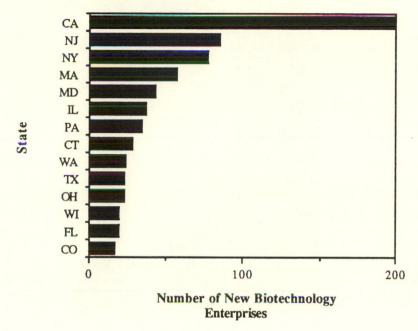

Number of New Biotechnology
Enterprises

public. The first major company to issue public stock was Genentech on October 14, 1980. The initial offering was one million shares at $35 share. Between March and July 1983, NBFs raised nearly $450 million in public offerings.[43]

DEMOGRAPHICS OF THE BIOTECH INDUSTRY

In the late 1970s when there was an explosion of NBFs, almost all of the expertise in genetic engineering techniques was at the universities. It was natural therefore to find the new firms locating in close proximity to major university complexes.[44] Many of these firms were started by university faculty who wished to retain their professorships while also participating in the development of a company. This latter phenomenon proved to have a major impact on the biological sciences throughout American universities and will be discussed more fully in Chapter 4. Figure 2.3 shows the 14 states with the most new biotechnology enterprises (NBEs) between 1973 and 1987. NBEs include new biotechnology firms (NBFs) and new divisions of long established companies (LECs). Figure 2.4 shows the 10 cities with the most NBEs for the same period.

Half of the biotechnology companies are located in just five states. California and Massachusetts are among the leaders in developing the new industry. This came as no surprise in that the preeminent institutions of molecular genetics are

Figure 2.4
Cities Leading in the Formation of NBEs, 1973–1987

City

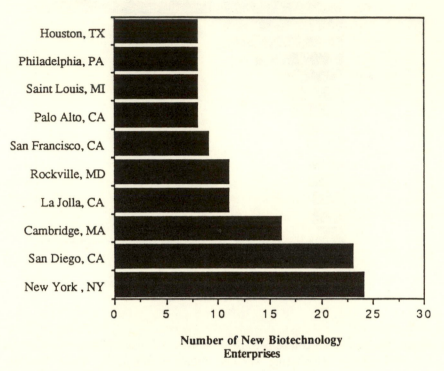

Number of New Biotechnology
Enterprises

located in both states (i.e., Harvard, MIT, Stanford, and U.C. Davis). The California Chamber of Commerce engaged in an advertising campaign to attract more firms to that state. One of their ads implored:

Come to California and bring your genes. Head straight for the best climate in the country. Especially for biotechnology. It's all happening right here; the place where the biotech industry was born. . . . Almost one third of the nation's biotech firms are located here. Nearly 20% of all R&D labs are here, too. . . . In fact, our schools have produced more experts in microbiology, genetics and engineering than any other state.

By 1988, two thirds of the states provided direct support to biotechnology programs. In fiscal year 1986 this support ranged from Utah's $110,000 to New York's multiyear appropriation of $34.3 million.[45]

Notwithstanding the rapid development of a new process industry in the context of highly publicized national debates over safety, most states resisted the passage of new legislation, favoring instead strong promotional campaigns for developing the new industry. According to a 1987 survey carried out by the Wisconsin Department of Natural Resources, only 9 out of 50 states were either developing or considering legislation to regulate any aspect of biotechnology. By December 1987 five states (Hawaii, Rhode Island, Michigan, New York, and Oregon) had legislation, resolutions, or rules to govern laboratory, greenhouse, or outdoor release of genetically engineered organisms.[46] By contrast, the OTA reported in 1988 that 33 states were engaged in some form of promotion of biotechnology.[47]

The desire of new firms to locate near the gene cloners exceeded even fears of burdensome local regulations. In Cambridge, Massachusetts, a city with the first rDNA ordinance in the United States, genetic engineering firms are required to have a license. They are overseen by a municipal biohazards committee that implements local health and safety requirements in addition to the provisions of the National Institutes of Health's rDNA guidelines. Despite the protestations of the local biotech industry against a city ordinance, many small firms located in Cambridge after passage of the law. Because of their proximity to Harvard, MIT, and the new Whitehead Institute, firms had access to university libraries, seminars, faculty consultants, and graduate students who could easily vault to and from the academic and commercial laboratories.

STAGES OF MATURATION

Kenney's book *Biotechnology: The University-Industrial Complex* provides a detailed analysis of the birth of the biotechnology industry.[48] Two other important sources on the industry's early development are the 1984 and 1988 publications of the Office of Technology Assessment, *Commercial Biotechnology* and *New Developments in Biotechnology* respectively.[49]

Within the first 12 years, the new biotechnology industry experienced a three-phased development. The first phase occurred approximately between 1976 and 1979 and consisted of the rapid organization of venture capital firms (see Figure 2.2). The second phase spans the period 1979 to 1983 and is distinguished by the efforts of the NBFs toward seeking financial independence and long-term stability. It was also a period when established companies began funding external projects and negotiating agreements with the NBFs. The third phase, starting in the late 1970s and continuing through the next decade, is denoted by the entrance of multinationals into biotechnology in the form of large-scale investments, intramural research, and equity holdings of NBFs.

A fourth phase of industrial development has already begun in the early 1990s, consisting of a winnowing out of R&D firms, a narrowing of the competitive field, and consolidation through mergers or buyouts of small firms by large corporations.

The commercial significance of gene-splicing became evident at the same time the scientific discoveries were being reported in scholarly journals. Ironically, public attention over the risks of genetic engineering brought additional media and investor interest in rDNA products.

By 1978 the media had become a captive audience to genetic engineering publicity. Every scientific finding was treated as a major technological break-through. Announcements of new discoveries in the major dailies were accompanied by predictions of cures for diseases and major agricultural applications. This publicity was important to a nascent industry that sought the confidence of the investment community and the general public that this was a safe blue-chip industry with enormous growth potential.

In the first phase of maturation, academic scientists were central figures in setting up the new firms. Many who played a principal role in establishing NBFs held on to their academic positions, a distinct difference between biotechnology and the microelectronics industry. In the latter case people left the university to start their own firms.

A few scientists like Harvard's Walter Gilbert relinquished their university positions when they were asked to choose between serving the company in a managerial capacity or holding their tenured professorial position. Harvard was one of a handful of schools with an institutional policy on faculty entrepreneurship. Gilbert gave up his position while heading Biogen for several years. When he was replaced as the president of the company, he was welcomed back to chair Harvard's biology department.

During the early stages of the new biotechnology industry, much of the R&D support came from venture capitalists and contract research from established firms. Research was organized around specific products. This was a period of intense competition among NBFs. A firm could not hope to succeed if it spent three to five years on development of a product only to learn that a competitor received federal approval first.

For the first decade, the strongest commercial activity of the NBFs was in pharmaceuticals. In a survey of 296 NBFs and 53 LECs, 21 percent and 26 percent respectively placed a primary focus on human therapeutics. Ironically, pharmaceuticals were among the most difficult products to bring to market. The U.S. Food and Drug Administration treated each rDNA product as a new drug, regardless of whether there was already an approved chemical equivalent on the market. The logic behind the ruling was that the new manufacturing process, microbial synthesis, might introduce different impurities. Thus, the FDA required the full panoply of tests.

NBFs were strongly dependent on larger firms for financing research and development. What they sold was their intellectual labor (i.e., their scientific skills for creating prototype organisms). Biotechnology is a highly knowledge-intensive industry. An OTA sponsored survey (circa 1987) revealed that of 35,900 jobs in biotechnology, 18,600 were for scientists and engineers.[50]

Many of the first NBFs (e.g., Genentech, Genex, Cetus) financed their own proprietary research by providing large established U.S. and foreign companies with research services for initial product development or by entering into licensing agreements with such companies that would result in future product royalty income.[51]

Table 2.5
R&D Research in Biotechnology Compared to Total for Large Corporations in 1982 (in millions)

	Total R&D	Biotech R&D	% of Total
Du Pont	750	120	16
Monsanto	220	62	28
Eli Lilly	267	60	22
Schering-Plough	127	60	47
Hoffman-Laroche	390	59	15

Source: Zsolt Harsanyi, *Investment Strategies in Biotechnology: The Race to Commercialization* (New York: EIC/Intelligence Pub. Co., 1984), 16.

The second phase of maturation in the biotechnology industry started around 1979. The NBFs began seeking independence from established firms by identifying their own products and defining and developing their own markets. Established companies gained enough confidence in the new genetic technologies that by around 1981 they began to develop in-house biotechnology R&D programs. Within a few years, biotechnology became a large part of the R&D budget of some major corporations (see Table 2.5). The amount of capital devoted to in-house biotechnology research increased dramatically in 1982.

An important indicator of this phase for NBFs is the rapid growth of public stock offerings between 1982 and 1983 from approximately $180 million in 1982 to nearly $600 million in 1983.[52] Another barometer of the effort by small firms to gain independence by manufacturing and marketing their own products is the decline in research contracts to NBFs awarded by companies between 1982 and 1983. But the path to independence for NBFs met with unforeseen obstacles. Even for the well-endowed leaders, one critical turn could cause financial havoc.

The Genex Corporation of Rockville, Maryland, was considered among a small elite group of the NBF leaders. It started its path to independence by manufacturing an enzyme-based drain cleaner called Proto which did not do well. Its big opportunity came when the company signed a contract with G. D. Searle to manufacture the amino acid L-phenylalanine, which was a key ingredient in the low-calorie sweetener aspartame. The contract provided Genex with 87 percent of the total revenue. The company purchased a plant in Paducah, Kentucky, for the manufacture of the amino acid. Then, in a decision that created shock waves through the company, Searle did not renew the contract.

On October 31, 1985, Genex announced that the company would undergo a complete reorganization. Its top management resigned, the work force was trimmed by more than half, and the Paducah plant was shut down. Genex had been put in a precarious position when a competitor was able to negotiate a better deal with its principal contractor for producing the amino acid. Obviously, one safeguard against the vulnerability of contract loss is for NBFs to manufacture products they can market directly. Another safeguard against contract uncer-

Figure 2.5
Aggregate Equity Investments in New Biotechnology Firms by Long Established
Companies, 1977–1983

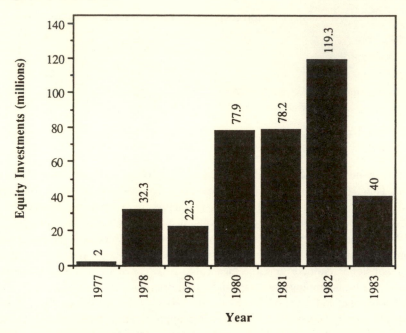

Year

Source: Office of Technology Assessment, *Commercial Biotechnology* (Washington, DC: U.S.
    Government Printing Office, 1984), 101.

tainties in the manufacturing end of biotechnology is an equity relationship
between the NBF and the established corporation.

In the late 1970s there was a rapid growth in joint ventures between NBFs
and major companies. The joint venture is a type of association (equity investment) that falls short of a merger. The two entities pool their resources for an
agreed-upon purpose, such as a particular product development. By 1982 the
equity investments in NBFs had peaked at nearly $120 million. In 1983, however,
equity investments decreased precipitously as large corporations turned to in-
house R&D programs (see Figure 2.5). Some equity investments in NBFs by
LECs have resulted in the formation of new firms. For example, Genentech and
Corning Glass begot Genencor.

NBFs have a survival incentive to innovate and market quickly. LECs can
afford to be more cautious and must since they already have an existing product
line: "A pharmaceutical firm with a vested interest in symptomatic treatment of
colds may have little incentive to develop a vaccine against the cold-causing
viruses, since it would diminish the company's sale of decongestants."[53] LECs
have a twofold interest in biotechnology: first, they seek to protect their established market share; second, they seek to develop new markets. Plants genetically

engineered to fix nitrogen may reduce the demand for chemical fertilizer. Likewise, plants engineered with built-in pesticide control could disrupt a multibillion dollar chemical pesticide industry. Chemical industry giants like DuPont and Monsanto have a substantial market to protect.

Both NBFs and LECs have something to gain by creating joint ventures. Genentech produced human growth hormone by recombinant DNA techniques. However, Eli Lilly dominated the animal insulin market in the United States. It had the name, the confidence of the medical community, and control over the distribution network. It is very unlikely that rDNA-derived human insulin would displace animal insulin overnight. Physicians were not likely to advise diabetics who were doing well on bovine insulin to suddenly shift over to human insulin. The collaborative venture between Eli Lilly and Genentech reduced commercial risks for both companies. Without Eli Lilly, Genentech would have had considerably more difficulty in getting Humulin (human insulin) to the market and in capturing a distribution network. On the other hand, a human insulin product was a great threat to Eli Lilly's substantial market share of animal insulins.

The second phase in the maturation of the biotechnology industry gave the NBFs only relative independence. Collaboration with the LECs enabled the NBFs to enter the product side of the industry, a large step beyond contract research. But the markets and production facilities were controlled by the LECs, who also had the burden of meeting regulatory requirements. Notably, the first few products submitted for regulatory review met considerable scrutiny (for example, see Chapter 7).

The third phase of maturation in the biotechnology industry began when the LECs began absorbing small successful firms, first as equity investments and then as buyouts. The less successful firms typically lost their funding sources or continued in a contract mode providing support to LECs.

Unlike the electronics and computer industries, there are limited military uses for biotechnology. A few of the R&D firms have taken on military contracts, but the long-term potential for defense industry support of biotechnology is not promising. The Biological Weapons Convention (1972), which restricts the production and stockpiling of toxins and viral and microbial pathogens, limits the development of a biological weapons industry. The Department of Defense (DOD) continues to fund biological research on the genetics of pathogenicity, vaccines, prophylactic responses to pathogens, and the potential uses of rDNA research on the putative assumption that said research will contribute to a defense against future uses of biological weapons (BW).

Notwithstanding the BW convention and the widely held view that biological weapons are not strategically important or militarily useful, the DOD expended nearly $119 million in biotechnology research for FY 1987 ($61 million in basic and $58 million in applied research). This was roughly divided between intramural (inside the agency) and extramural (outside the agency) research. To some observers this figure, a 28 percent increase in the DOD's biological research since FY 1985, suggests a growing militarization of the biological sciences.[54]

The biotechnology industry has been fully integrated into the U.S. and world economy. Capitalization of the industry has been based on extravagant promises of multibillion dollar markets. Some of these promises have been realized; many have been exaggerations by an industry that has been apprehensive about public perceptions of genetic engineering. Unlike previous technological revolutions (plastics and microelectronics) that have basically sold themselves, biotechnology has met various forms of public opposition, particularly in the second strongest sector—agriculture.

The rapid development and mainstreaming of biotechnology is in part a result of the cooperation of the academic, government, and corporate sectors. It has been an industry of fashion for the 1980s. Even metaphors like "designer genes" capitalized on the idea of refashioning nature. It is an industry that drew heavily from public concern over dread diseases like AIDS and cancer and annoyance diseases like herpes and colds. It is an industry that delighted in the prospect of manufacturing transmogrified animals. The coordinator of biotechnology for Iowa State University stated the new agenda rather bluntly: "This is the first time man has been able to manipulate plants or animals in a way nature hasn't provided. It's really remarkable, because now we can take something out of a banana [a gene] and put it into a cow."[55]

Applied molecular genetics represents a new frontier. And like the draining of the English fens, the unleashing of nuclear energy, the taming of the American West, and the damming of the world's great rivers, it is subject to the excesses of its zealots. Technological revolutions invariably create second-order effects that arouse political controversy and stimulate philosophical and legal debates. The next two chapters examine the social and ethical issues associated with the patenting of life-forms and university-industry relations.

3

Patenting Hybrids, Chimeras, and Other Oddities

How can anybody say [patenting animals] is unethical or wrong?
> Donald J. Quigg, Assistant Secretary of Commerce and
> U.S. Commissioner of Patents, April 13, 1988[1]

Future generations will look back and shake their heads in disbelief at Government support of this attack upon the integrity of life.
> Tom Regan, President, Culture and Animals Foundation, 1988[2]

Scientific innovation is only one of the elements required for an industrial revolution. The creation of legal, social, and economic structures that support the innovations is an essential component of technological change. Technical innovations and societal structures co-evolve. Thus new agricultural technologies were brought into use by the creation of federally sponsored agricultural field stations and publicly financed agricultural colleges. The commercial uses of nuclear power became viable in the United States when Congress enacted the Price-Anderson Act that limited total claims against a single nuclear facility to $560 million in the event of a nuclear accident. Tax credits for energy-saving devices were directly responsible for the commercial profitability of many new conservation-oriented technologies.

The biotechnology industry began with sets of innovative plans for new pharmaceuticals and agricultural products. Some of these plans fit nicely into the existing institutional structures. For others, technology transfer from the laboratory to the manufacturing plant required changes, or at the very least, a clearer articulation of the legal status of biotechnology products in two principal areas: patentability and regulation. Changes in the regulatory status of products will be discussed in Part III. In this chapter I shall explore the evolving interpretations of patent rights prompted by the advances in genetic technologies.

A PATENTED MAMMAL

Patenting is a right guaranteed by the U.S. Constitution. Nevertheless, there is no evidence that the framers of that provision (Article 1, Section 8) had ever anticipated it would be applied to living things. Sixty years ago plant patenting was an issue before the legislature. In the face of the explosive developments in biotechnology, Congress and the courts had to revisit the question: Is life patentable? From plants, the question turned to microbes and animals. What if any limitations should be placed on patenting living things? By issuing a patent for a "new" living product, is the government explicitly giving its support to the technology? Are there any religious principles or social and ethical norms that are inconsistent with patenting living entities? Will the patenting of life-forms in general and animals in particular have any undesirable social, economic, or environmental consequences? These and other queries brought crosscurrents of discourse among animal rights activists, patent lawyers, farmers, and molecular biologists.

On April 13, 1988, the major dailies ran a story on the first patent (#4,736,8666) issued for a higher life-form. The patent was assigned to Harvard College and issued to Harvard Medical School geneticist Philip Leder and Timothy A. Stewart of San Francisco. Leder appeared prominently in a photograph with his finger pointing to an enlarged image of a rather exotic mouse. The mouse was the object of a genetic engineering experiment that left it with the distinguishing characteristics of expressing a cancer-causing gene in its mammary tissue and being hypersensitive to toxic substances. The United States had taken the lead among the patent-granting nations in establishing patent protection for genetically altered mammals.

Public reaction to the announcement of the patent was relatively mild. Criticism came primarily from animal rights groups, some environmentalists, and farmer associations. Two bills were filed by members of Congress requesting a moratorium on animal patents until the issue had been thoroughly examined. There were several different reasons among those opposing the decision to patent the mouse.

Some individuals wondered whether it was merely a coincidence that the Patents and Trademarks Office (PTO) chose the "cancer-mouse" as the first patentable mammal and not some chimeric farm creature. At the time there were 21 patents for genetically altered animals pending before the PTO. But among the other possibilities, the research mouse would most likely appeal to people's humanitarian concerns and fear of cancer. This onco-mouse was the latest product of scientific ingenuity in a long and often disillusioning war against breast cancer.

Applying techniques of microinjection of DNA, Harvard scientists inserted into mouse eggs a gene they had isolated from other strains of mice that can produce cancer in many mammals, including humans. The gene was modified in such a way that multiple copies of it function in the mouse's mammary tissue.

An unusually high proportion of these mice with the oncogene develop breast cancer.

This new breed of mouse could now serve as a model system for studying the development of cancer. It has several benefits over the standard laboratory strain of germ-free mice. First, it is more sensitive to cancer-causing agents. Second, it will allow scientists to study the role of oncogenes and environmental factors in the production of cancer of the mammary tissue. Third, scientists claimed it offered a promising model to test new drugs and therapies to treat breast cancer. The importance of the patent—a device to protect financial risk in the development of new technologies—was closely tied to the significance of the mouse as a research tool to help conquer a dread disease.

The mouse is called transgenic because it was produced by transplanting genes from one strain or species, or even kingdom, to another. Animals created in this way may carry the transplanted genes to every cell and pass them on to offspring. The statement of the patent claim is for a "transgenic non-human eukaryotic animal [preferably a rodent such as a mouse] whose germ cells and somatic cells contain an activated oncogene sequence introduced into the animal, or an ancestor of the animal, at an embryonic stage." (The germ cells are the reproductive cells; the somatic cells represent all other cells that make up the organism.) The patent covers not only the strain of mice, but any mammal (except humans) that has had any of a variety of cancer genes introduced into its embryo—one or more cells, but usually not later than the eight-cell stage. Both the process of production and the product are covered by the patent.

Before examining some of the social and ethical problems associated with patenting transgenic animals, I shall give a brief historical overview of the U.S. experience with patenting living matter.

THE LEGAL HISTORY OF PATENTING LIVING THINGS

The U.S. Constitution, ratified in 1787, includes a clause that established the foundation of all subsequent patent and copyright law.

Congress shall have power . . . to promote the progress of science and useful arts by securing for limited times to authors and inventors the exclusive right to their respective writings and discoveries.[3]

This provision reflects the values of men like Thomas Jefferson, himself a distinguished architect, inventor, and political philosopher, who sought in the Constitution a means to foster the applied sciences in an effort to build a new political and technological world. The framers of the Constitution introduced the concept of intellectual property rights in contradistinction to civil rights and the rights of property ownership.

Two and a half years after the Constitution was signed, Congress passed the

Patent Act of 1790 (amended in 1793) that provided for patents to be issued to any person who has "invented or discovered any new and useful art, machine, manufacture, or composition of matter, or any improvement of these not before known or used." By the 1850s, America was embarking on a golden age of technology. Europe had its high science, but America was spawning artisans, tinkerers, and engineers. Discoveries came from home laboratories and shops. Invention was a cottage industry and the patent awards accelerated at an exponential rate between 1840 and 1870 (1,600 patents were awarded by 1840 and 80,000 by 1868). In 1878, patent number 200,521 was awarded to Thomas Edison for the incandescent lamp; in 1889 Charles Hall was awarded patent number 400,665 for the manufacture of aluminum; and in 1906 Wilbur and Orville Wright received patent number 821,393 for their flying machine. Soon after the first decade of the twentieth century, America awarded its millionth patent—all for mechanical objects, chemical substances, and some processes involving bacterial agents (e.g., Louis Pasteur received a patent for purified yeast in 1873).

In the 1920s plant breeders sought the same opportunity to reap the benefits of the patent system as those innovators in mechanical arts. Congress responded by passing the Plant Patent Act of 1930 and thereby removed the distinction between plant developers and industrial inventors. The act extended the definition of patentable material to include certain varieties of asexually reproduced plants (those propagated by cuttings, grafting, and budding, but not by seeds). Supporting reports from the House and Senate stated that there was no difference between plant grafters and the chemists who create new compositions of matter. As a measure of the public support or indifference, the 1930 Plant Patent Act was passed within three months of the time the bills were first introduced.

The next 40 years brought extensive innovation in plant breeding in conjunction with the rapid rise of pesticide and herbicide use in agriculture. In the dawn of the Green Revolution—the term that signifies the new capital-intensive approach to agriculture along with innovations in hybrid seed development—Congress extended plant variety protection to any novel strain of sexually reproduced plants (other than fungi, bacteria, or first-generation hybrids) by enacting the Plant Variety Protection Act of 1970. This act made possible the patenting of major food crops that were developed by classical hybridization techniques, an arduous process of crossing, screening, and selecting new strains of seeds for desirable traits. Within ten years more than 200 varieties were registered with the Patent and Trademark Office.[4]

Legal scholars have frequently maintained that passage of the 1930 and 1970 plant patent acts is evidence that the language "manufacture" and "composition of matter" introduced by an earlier Congress did not apply to "living things." Otherwise, it is argued, Congress would not have passed specialized legislation for plant patenting. This issue has divided the legal community. Other legal scholars contend that Congress took action with regard to plants to set the parameters that define novelty, not to decide, once and for all, whether living

Table 3.1
Single-Celled Organisms Awarded Patents Prior to 1930

Date	Inventor	Patent Number	Claims
1873	Pasteur	141,072	Purified yeast
1908	Coates	899,155	Ground vegetable or animal matter inoculated with bacteria
1910	Collett	952,418	Bacteria mixed with cocoa
1914	Earp-Thomas	1,099,121	Sterilized soil inoculated with bacteria
1914	Odle	1,120,330	Food product containing lactic acid bacilli
1916	Palma	1,178,941	Bacteria combined with nitrates
1918	Harris et al.	1,260,899	Lactic acid bacteria and inert material
1923	Stoltz	1,442,239	Nitrifying bacteria combined with calcium carbonate, phosphate rock, and humus
1923	Whitmore	1,457,097	Microorganisms in vegetable oil
1925	Reichel et al.	1,540,951	Lactobacilli and culture media

Source: Bureau of National Affairs, ''Briefs Filed in the Chakrabarty Case,'' *Patent, Trademark and Copyright Journal*, no. 465 (Feb. 1980): D-14.

things are patentable material. They correctly observe that well before patents were awarded to plants, the Patent and Trademark Office approved patents to single-celled organisms in the context of a process of manufacture (see Table 3.1). This is cited as evidence that the living/nonliving distinction has not been the crucial factor for the PTO in issuing patents.

Congress has never chosen to pass special legislation to address the patentability of single-celled organisms. Up until 1980, single-celled organisms were issued patents for their contributing role to a process. In that year the Supreme Court considered a patent claim for a bacterium that established a new precedent for patenting living things.

While working for the General Electric Company, Ananda Chakrabarty developed a bacterium that degraded the components of crude oil. His patent application included the process of making the microbe, a method of dispersal, and the organism itself. The U.S. Patent and Trademark Office accepted the first two claims but denied the third. The main objection by the PTO was that the rearrangement of genetic materials and modification of life-forms does not qualify resulting life-forms as a product of manufacture or a new composition of matter. The patent examiner denied the patent on the grounds that (1) microorganisms are products of nature and (2) as living things they are not patentable subject matter.

The PTO Board of Appeals, however, reversed the examiner on the first claim, contending that the modified bacteria are not naturally occurring and therefore qualify as products of manufacture. But the appeals board accepted the general exclusion that bacteria are not patentable as living entities whether or not they are natural or man-made. The next higher court, the Court of Customs and Patent Appeals (CCPA), ruled that the living status of the entity is not grounds to exclude a patent claim.

The case was appealed to the Supreme Court, which by a 5–4 vote on June 16, 1980, decided in favor of the patent award for the oil-degrading microbe. Chief Justice Warren Burger, speaking for the majority in support of the CCPA interpretation, held that new constellations of living matter are no less products of manufacture than rearrangements of inert substances. The relevant distinction in patenting, according to Burger, is not whether the entity is living or dead but whether it is a product of nature or a human invention. The court concluded that Congress intended patentable subject matter to "include anything under the sun that is made by man."

Prior to the Chakrabarty case, there was no confusion about patenting a single-celled organism when it was part of a process. Numerous microorganisms were patented by the food fermentation and pharmaceutical industries. But in this decision, the oil-degrading microbe was issued a patent *sui generis*, regardless of how it might be used. Under the new interpretation, the patent holder had a much wider range of protection.

Building on the court's decision, the PTO issued a policy statement in April 1987 indicating that it considers nonnaturally occurring nonhuman multicellular living organisms, including animals, to be patentable subject matter within the scope of the patent laws. During the same month the Board of Patent Appeals and Interferences upheld the claim that polyploid oysters (oysters produced by making multiple copies of genes) are nonnaturally occurring manufactures and thus patentable material. This novel oyster was created by hydrostatis pressure, a process which did not involve biotechnology. Finally on April 12, 1988, the first patent award for a mammal—the cancer mouse—was announced.

In reviewing this brief history, it is noteworthy that the decisions to patent living things moved to a progressively narrower forum: from legislative decisions following hearings and public debate, to the Supreme Court's 5–4 vote, to an administrative action by the assistant secretary of commerce under the Reagan administration (who also served as the commissioner of patents and trademarks). A narrowing of public involvement in the decision occurred as the ethical dimensions of patenting became more complex.

Federal policy has also played an increasingly significant role in fostering patent activity within industry and universities. Congress passed the Government Patent Policy Act of 1980 (PL 96–517), with amendments in 1984 (PL 98–260), that set up uniform standards for giving small firms and universities patent rights to discoveries that resulted from federally sponsored research. A presidential memorandum in 1983 broadened those policies to include large businesses and contractors who receive federal support. The Office of Technology Assessment reported that five years following passage of the 1980 patent legislation, patent applications for human biologicals by universities and hospitals had increased more than 300 percent over the preceding five-year period.[5] Under the Reagan administration, universities received strong federal incentives to patent their discoveries and form partnerships with industry. Biotechnology was viewed as being among a few emerging areas for U.S. global industrial leadership. The climate was not favorable for protracted debates over the value-laden issues associated with the patenting of animals, since such debates were viewed as a threat to research and development, technological innovation, and industrial efficiency. Nevertheless, debates over patenting animals did take place, although in limited contexts among well-defined stakeholders. Some of these debates were truly not about patenting per se, but that was the convenient rubric for raising other concerns.

ETHICAL AND SOCIAL POLICY ISSUES

Patenting has never been a subject of strong public passion. Some political debate took place when the plant protection acts were being considered. Mainly, it involved plant breeders, garden societies, and horticulturists. When the acts were enacted the controversy subsided quickly. The Chakrabarty decision drew a wider response from public interest groups since patenting, genetic engineering, eugenics, and equity in agriculture were all intertwined. When the Supreme Court issued its decision that the oil-degrading microbe was patentable material, public criticism subsided and Congress saw no reason to address the issue beyond the airing that took place during the hearings.

The decision to patent a mammal brought many of the advocacy groups that opposed the patented bacterium into the latest policy fray. It also attracted another formidable constituency, animal rights groups. The concept of a patented animal signaled to these groups that society was regressing to an extreme Cartesian view of animals as soulless, unfeeling creatures that may be treated like machine parts.

The announcement by the PTO of the patent award for the cancer mouse precipitated a host of social and ethical arguments for and against patenting animals. Several of these arguments will be discussed in subsequent sections. For purposes of clarification, it is useful to distinguish arguments that focus on the application of the technology from those that focus on the patentability issue.

Table 3.2 represents four cases, each with a distinctive and consistent viewpoint. For example, one may support the patenting of animals, i.e., either as part of a process or as products of traditional breeding, but oppose the genetic modification of animals. Individuals who hold such a view may be inclined to view the patenting issue through a legal framework and the genetic engineering issue through an ethical framework.

It is also not uncommon for someone to favor the genetic manipulation of animals while opposing the patenting of them. There are legal, financial, and philosophical reasons why people oppose patenting and these may not carry over

Table 3.2
Distinction between the Use of Gene Technology on Animals and the Patentability of Animals

Use of Gene Technology on Animals	Patentability of Animals
Agree	Agree
Agree	Disagree
Disagree	Agree
Disagree	Disagree

to the use of the technology. A farmer may delight in having a genetically engineered pig that is leaner and tenderer than its natural counterpart. But the prospect of the pig being patented may present special economic burdens that farmers deem undesirable.

Cases where there is agreement or disagreement for both the use of the technology and the patentability of animals draw attention to the political extremes (factory-farm advocate versus deep ecologist, e.g., someone who believes humans have no moral claim over animals).

Having separated the patenting issue from the application of transgenic technologies to animals, however, I should discuss an important sense in which they are interrelated. Patenting a technique may have an effect on its use. Even if this is not actually the case, people may believe that such an effect exists. Opposition to patenting may be based on a real or imagined relationship between the class of patented objects and the advancement of a particular class of technologies. It is an open empirical question whether there is any causal relationship between the decision to patent animals and the spread of transgenic animal technology. Those frustrated about the lack of control over the technology may seek to block animal patenting on the assumption that without it as an incentive to industry, the program to modify the genetic architecture of animal species may lose its economic incentive.

THE CONTROVERSY OVER PATENTING ANIMALS

Animals have a moral status in our society that plants, microorganisms, and oysters do not possess. Moreover, the political activism of animal rights organizations has intensified in recent years and public policies toward the treatment of animals are still evolving. By adopting a narrow legalistic approach to the issue of animal patenting, the opportunity to address the nascent ethical and social concerns that are unique to this context is lost. Congress, more than the courts or a patent administrator, can hear these claims. From the standpoint of the decision process, there is excellent reason for Congress to resolve the legal wrangles over the correct interpretation of patent statutes. It is the only body that can legitimately set public policy for patented living material. Congress can balance the interests of breeders, animal owners, and farmers with those of the rest of society in addressing such questions as: Can animals genetically engineered to develop human organs be patented? Should there be a research exemption for patented animals? If a farmer purchases a patented cow, does he have exclusive rights to its progeny? When animals are patented should the patent holder be required to submit a sample of the animal for deposit in a facility similar to the cell culture repository? Without congressional entry into this field the courts or the patent administrators will continue to set public policy and establish legal conventions and ethical norms.

The same year the patent was awarded for the onco-mouse, congressional movement on the patenting issue intensified with hearings, resolutions, and the introduction of several bills.[6] In 1987, the Senate adopted a moratorium on animal patents that was appended to an appropriations bill. The rider was subsequently dropped after a House-Senate conference. Between 1987 and 1988 there were bills introduced into both houses to prohibit patenting of genetically altered or modified animals. No action was taken on these initiatives. Perhaps the most serious attempt to date at legislation is the Transgenic Animal Reform Act introduced by Robert W. Kastenmeier (D–WI) on June 30, 1988, and passed by the House later that year. This bill accepted the patentability of nonhuman animals but provided for an exemption to farmers who reproduce the animal and to researchers who wish to experiment on the animal for purposes of improving it or developing another strain. This act is modeled closely on the Plant Variety Protection Act, which has similar provisions. The report of Kastenmeier's subcommittee argued for the farmer's exemption on pragmatic grounds. "By the very nature of their occupation farmers will use patented animals on their own farms for reproductive purposes. Because this is an area in which contractual solutions are unlikely to be effective a statutory exemption is necessary."[7]

Patent law is built on a philosophy that patent protection must nourish innovation and not stifle it. Congress listened to many stakeholders, some of whom viewed animal patenting as a new and radical policy change. Others, including the consensus of patent attorneys, treated animal patents as an incremental and logical step in a long historical process.

Many of the criticisms that have been directed at animal patenting cite undesirable consequences such as narrowing of species diversity, monopoly ownership of animal species, increased suffering of animals, adverse impacts on Third World countries, and a deterioration in social attitudes toward animal life. Each of these arguments is premised on a causal relationship between the patenting decision and one or more adverse outcomes. It is very difficult to demonstrate in some conclusive way through empirical methods the causal effects postulated. But there are some analyses that make persuasive arguments regardless of their relevancy to the constitutional question of patent rights.

For example, after patent protection was extended to seeds, major corporations invested heavily in plant breeding. They sought large markets and uniform patent-protected products (germ plasm). There is some historical evidence to support the claim that as an industry becomes more heavily concentrated, it follows a path toward greater product homogeneity in order to capitalize on economies of scale. There is also evidence of a trend toward the use of fewer seed varieties in crop production. But if there is a narrowing of the seed base is that the result of patenting or would that have happened regardless? And what are the tradeoffs in having multinationals control global seed markets with fewer high-yield varieties? These issues remain hotly debated and divide agricultural analysts along ideological lines. OTA reviewed the claims and counterclaims about the relationship between seed patents and the availability of germ plasm and concluded that ''any information regarding the impact of intellectual property protection of plants on germ plasm is largely anecdotal.''[8]

I do not intend to explore the empirical bases of these claims in this work. However, structural arguments of the consequentialist type deserve more careful examination. If the patenting of animals is a right, then why should the actual or possible use of this ''novel object of manufacture'' be relevant to the exercise of that right? The Patent and Trademark Office has never been required to undertake a social or environmental assessment of a patent application. Moreover, Congress has thus far not limited patenting on the grounds that certain applications of the patentable materials might be socially undesirable. That would exclude many patentable items whose socially redeeming functions are questionable. When Congress initially restricted plant patents to asexually propagated varieties in its 1930 law, it did so on pragmatic grounds because it believed sexually reproducing plants could not be replicated ''true-to-type.''

Thus, even if it can be demonstrated that there are some adverse social or economic consequences to patenting animals, there is no tradition in the history of patent policy and patent law for restricting such awards on these grounds.

The Reproductions of a Patented Animal

When a patent is awarded for some process or ''object of manufacture,'' the patent holder can set licensing fees for commercial uses of the product or process, though licensing fees may not be so restrictive that they virtually block the use of the invention. Moreover, a patent must disclose enough detail to enable an informed individual to replicate the invention.

Patent law and patent administration are fields of considerable specialization. They deal with highly technical and sometimes philosophical issues as: What constitutes novelty? When is the modification of an invention considered patent infringement? What changes in a natural object qualify it as patentable material?

These issues are difficult enough when applied to mechanical objects (non-living entities). When patents are issued for living animals, another layer of complexity is introduced. The Patent and Trademark Office must now contend with the problems of reproduction, progeny, and claims that the patented entity has rights of its own.

The holders of the mouse patent established as their claim any nonhuman mammal that has an oncogene sequence introduced into its genome during its embryonic stage. Presumably, the novel genetic sequence is not found naturally in the mammalian genome, otherwise the animal would not be patentable. Some well-defined and replicable genetic modification of the embryo of the animal qualifies the animal for a patent.

Let us suppose that someone purchases a non-patented animal (such as a domestic or farm animal). The animal is owned by the purchaser, who may resell it, have it neutered or declawed, or if it is a farm animal, have it artificially inseminated or slaughtered. The owner is bound only by laws regulating the humane treatment of animals. Now suppose the animal one purchases is from a patented strain. Under such conditions the owner may have unrestricted rights over the physical possession of the animal as a material living object, but only restricted rights over the animal's reproductive life, at least over the duration of the patent. The right to breed the animal without financial obligation to another party can no longer be assumed to be part of the rights of ownership, unless the PTO or Congress makes a special provision in this regard. This may be no less true for patented bacteria or oysters but it takes on a different dimension in mammalian species. For people who have spent a sizable sum of money for their dog, steer, or horse, it may seem strange that the right to breed the animal does not come with the ownership papers.

There is another peculiarity in extending the concept of patenting to animals. Following the PTO's decision on the onco-mouse, the fertilized egg of the mouse was genetically modified to qualify for a patent. The reproductive biology of the animal is wholly natural and presumably non-patentable. The human "manufacture" is at best only in the fertilized egg. When the animal exercises its natural reproductive functions, the inserted genetic sequence is replicated in its offspring. Why should the patent claim extend to the living products of a natural process of reproduction? If a typewriter is patented, its replication is fully controlled by manufacturing techniques. Human intervention takes place at every phase of development: design, manufacture, and reproduction. This is far from the case when the fertilized egg of an animal is modified.

Suppose someone purchases a female dog that has been bred with bright spots by genetic modification of the species ovum. The patent claim is for the spots. When the animal bears offspring, presumably some are spotted and some are unspotted. Under the current patent rules, the natural biological process that determines whether the genes for spots get transferred to the progeny will also determine the patent liability of the progeny. If the size or shape of the spots are diminished after one or more generations, does the patent holder have a smaller claim?

We might consider a patent claim for several traits like spots, eye color, and tail size. Some progeny may have all or part of the genetically programmed phenotypes. The right to patent animals will create all sorts of problems in adjudicating patent claims for progeny because in the reproductive life of the animal it is virtually impossible to isolate the roles of human intervention and natural process.

Without covering reproduction, an animal patent would have no advantage over traditional breeder's rights, since the patent holder would have legitimate licensing rights over the first purchase only. Thus, if patenting animals is to make financial sense, there is a compelling argument that the claim must cover the animal's natural reproductive biology.

Pragmatic Basis for the Farmer's Exemption

The livestock industry has lobbied forcefully to restrict the scope of the patent claim to the initial sale of an animal on the grounds that it would be impractical to extend the claim to the progeny. Beef cattle farmers exercise many options on the use of a particular animal during its stages of development. If progeny were included in the patent claim a tracking system would be necessary for identifying the traits under patent coverage that appear in multiple generations. Not all progeny from a patented animal will contain all the lucrative traits. Furthermore, some may have the genotype but not the phenotype.

It has been suggested that these problems can be solved by technical innovations in mass genetic screening. But even with such efficient methods to identify the transplanted genes in multiple generations, it would still be problematic whether the farmer was receiving the economic value of the genotype. If the patented animal was genetically engineered with several traits, within a few generations the traits would be redistributed among progeny and difficult to track. Moreover, cattle are often sold more than once and pass through ownership of several brokers. The problem of implementing a tracking system is exacerbated by the highly decentralized character of the beef cattle sector.[9]

One technological solution to the problem is to render the patented animal sterile. This is analogous to computer software that is usable but uncopyable. Another solution is to create a strain whose desired patented trait cannot be transmitted by natural reproduction. For example, suppose the sperm of the livestock with the desired genotype is patented and only the sperm that produces female progeny are sold to livestock farmers. The result of this system would be to remove some traditional breeding practices from the farmer's domain. Just as second and third-generation seed is not as productive as the hybrid strain, under this scenario, livestock farmers will constantly require new germ plasm and the progeny issue will be moot.

Patenting and Animal Abuse

Animal rights activists have established a high visibility on the issue of animal patenting. Some representatives of the movement claim that the right to patent animals will result in their greater abuse. The claim is based on the argument that patenting will foster more radical forms of genetic engineering, and that some of these forms will involve the creation of animal hybrids or chimeras in which the novel strains will suffer increased stress, physical pain, and hardship. By redesigning its molecular biology, it is argued, the physical being of the animal will be estranged from its species nature.

The creation of transgenic animals might result in "transmogrified cripples." These are animals whose organs and limbs do not contribute to an integrated morphology. Consider the case where an animal is genetically engineered with a reproportioned body (i.e., a pig with shorter legs, chickens with larger breasts, a cow with a larger udder). Radical reconfigurations of an animal's physiology may restrict the animal from expressing its species nature. For example, the transgenic anatomy of the animal may inhibit motion because of severe weight, alter digestion, or make reproduction extremely painful. Moreover, redesigning an animal's genetic makeup solely for the purpose of enhancing its commercial value negates the animal's sentient qualities and reduces it to a lifeless amalgam of protein molecules.

Citing the case of genetically engineered bovine growth hormone (BGH), Michael Fox of the Humane Society of the United States has argued that new agricultural technologies will increase the abuse of farm animals. The hormone is used to hyperstimulate cows to produce 20 to 40 percent more milk—a process that Fox contends induces more stress and suffering.[10] Following the path of biotechnology, Fox maintains, "animals, as we know them today, will cease to exist since their telos (intrinsic nature and ultimate purpose) will be wholly subjugated under the dominance of absolute materialism."[11]

Such examples of "designer animals" have begun to appear. Scientists at Beltsville, Maryland, created a boar by injecting cattle genes into fertilized embryos of a pig and implanting the embryos in the womb of a surrogate sow. One journalist gave the following description of the animal. "Its legs are thick and stubby with arthritis. Its head is overly broad. It is highly susceptible to disease, its hair sparse and its brow over grown."[12] Researchers thought they could make a leaner pig. Instead they ended up with an arthritic, lethargic beast with an overgrown skull and crossed eyes. Genetic engineering is not the only means of creating bizarre morphological structures. Through traditional methods of selective breeding, turkeys have been bred with breasts too large to permit them to mate.[13] The latter example has been used to discredit the argument that patenting genetically modified animals will foster unique consequences.

Bernard Rollin, who has written widely on animal rights, uses the Aristotelian term "telos" to circumscribe the animal's basic interests. He argues that those interests ought to be legally protected.

So the main moral challenge to those involved in genetic engineering of agricultural animals is to avoid modifying the animal for the sake of efficiency and productivity at the expense of the animal's happiness or satisfaction of its nature . . . it would be grossly immoral to use genetic engineering to change chickens into wingless, legless, and featherless creatures who could be hooked to food pumps and not waste energy.[14]

If patenting is viewed as having no relevance to the use of transgenic technology, the previous arguments lose their force. On the other hand, while patenting may not be a necessary condition to the creation of transgenic and chimeric animals, it is at least a stimulus. Whether or not animal patents continue to be awarded, there is a vigorous effort underway to create new strains of cows and pigs with lower fat content and higher nutritional value. This is a response to heightened consumer awareness resulting from health warnings to reduce high-fat diets. There are also economic and environmental incentives to improve the efficiency by which animals transform grains into protein.

It is unlikely that patenting will make an appreciable difference in the commercial efforts to genetically engineer animals. Appealing to humane treatment arguments to counter transgenic animal technology seems disingenuous to some observers in the context of the animal brutalization that already exists in factory farms. According to Baruch Brody:

We should, I believe, conclude that our regulations governing research on animals to minimize suffering should be strengthened to protect farm animals, birds, and rodents used in privately funded research. However, unless we are prepared to radically alter the entire way in which we relate to the animal kingdom, none of the articulated arguments offered by [the animal rights movement] should lead us to oppose the patenting of transgenic animals.[15]

Nevertheless, issuing patents further legitimates the commodification of animals that is already pervasive in agribusiness and extends it to research, pet breeding, and sports. Unless genetic engineering can create animals that do not feel stress, cannot experience pain, have no social needs, and possess no species requirements (in other words are inert entities), the legal reduction of animals to mere compositions of matter will never sit well with some segments of the population.

Most legal experts view the patenting of animals as an inevitable outcome of the Chakrabarty case. But in public policy terms there are significant differences between bacteria and mammals. People do not get excited about bacterial reproduction (unless the microbes are pathogenic) and it makes no sense to speak about abuse of bacteria. Neither insects nor bacterial species are protected by an endangered species act. When patents are finally awarded for domesticated animals and begin to interfere with the laissez-faire attitudes of animal breeders, the issue will trigger another set of policy debates that pit legal logic against social interests.

Final Observations

I draw several conclusions from the animal patenting debate. The law makes a critical distinction between the right to patent an invention and the right to manufacture and introduce it into commerce. The patent system is designed to keep these roles independent. Many patented inventions are so heavily regulated that they are not commercially viable. Some patent-protected inventions may be unethical to commercialize. But in the opinion of most patent experts, these are issues that transcend the patent system and should not be addressed within it.

The controversy over "patenting life" highlights some important traditions of American culture. This nation's appetite for technological innovation that spawned the first industrial revolution remains robust; the patent system serves as one of its prime incentives. Patent law has insulated the value of innovation from other political and social considerations. For those entrenched in the patent culture (bureaucrats, lawyers, and inventors), innovation has intrinsic worth independent of its medium of expression.

Another observation is that changes in the patent and regulatory systems are being driven by scientific conceptions of nature. Science has broken down the traditional boundaries between inert matter and living things. The legal and regulatory systems are beginning to accommodate to these changes at a pace that has perhaps left much of the public quite baffled. People are not fully acclimated to applying mechanistic concepts to life forms.

Finally, the issue of patenting living things has served as a social heuristic and provided a vehicle for discussions about the direction and control of technology. One segment of society benefits by restricting the debate to fine legal details. Other segments are served by recasting the debate into broader terms. The battle in the legal domain is merely a surrogate for another layer of questions about control over biotechnology and its differential impact on society. Once legal issues are resolved, other social heuristics arise providing for similar opportunities.

4

Science and Wall Street: Academic Entrepreneurship in Biology

To maintain the confidence and support of the public, the university must be seen to be unbiased in its expert judgments and committed primarily to the public interest through the pursuit of new knowledge and the cultivation of scholarship and education.
> Arnold Relman, Editor, *New England Journal of Medicine*[1]

There is nothing wrong with universities serving as a source of technical expertise for newly developing industries or even for mature industries. And that kind of symbiosis between industry and academic departments has been very good for both over history. . . .
> David Baltimore, Professor at MIT[2]

Commercialization finally caught up with academic biology as it had other pure sciences some years earlier. The industrial revolution in applied genetics was accompanied by changes in the relations between university scientists, their institutions, and the nascent biotechnology industry. With these changes came a new scientific ethos in the university. According to this ethos, the production and dissemination of knowledge is fully realized in an instrumental sense only when the knowledge is transferred to the industrial system. The newly conceived role of the scientist is to bring both the generation and application of knowledge to fruition. In this chapter I examine the partnerships that developed between the biological sciences in American universities and the emerging biotechnology industry. I also review the findings of those who have studied this phenomenon in an effort to gain a clearer understanding of how the commercialization of biology has affected the social mores of research and the boundaries between academic and industrial science.

THE MATRICULATION OF BIOLOGY

Traditionally, students majored in biology because they were in a premed program, they were interested in an environmental career such as conservation or ecology, or they possessed an intrinsic interest in the subject matter of living things. Biology was not generally considered a pathway to the industrial or corporate world. Outside of academic research, microbiologists could find employment in the food and pharmaceutical industries. Human and plant geneticists had very little professional opportunities in the private sector. Much of the plant genetics that spawned the Green Revolution was carried out in the land grant agricultural colleges or at federally funded or internationally supported institutes.

This situation changed rapidly within a few years after the discovery of recombinant DNA. Molecular geneticists began to realize that the knowledge they possessed was the key to a multibillion dollar industry. Biomedical science suddenly became swept up in a gold rush atmosphere. The prospectors were the new generation of gene sequencers and cloners.

Among many biomedical scientists, there was a strong propensity to find a place for themselves in the industrialization of biology. Some viewed this as a sign that their field had matured. Molecular geneticists could now stand alongside their colleagues in chemistry, physics, electronics, and computer sciences and say that their knowledge too, had important industrial payoffs. The commercialization of the medical sciences had already begun in the 1960s with the burgeoning of biomedical technologies in health care. However, for 25 years molecular biology remained a field of basic science. And when commercialization finally came, it was ushered in by those preeminent in the field. They were the role models for the new generation of molecular geneticists and biotechnologists.

There are two notable aspects to this phase in the industrialization of science. First, it was not prompted by military applications or defense funding. Second, it was accompanied by an unprecedented wave of media attention, controversy, and self-examination within the academic community. If, as some biologists proclaimed, molecular genetics was simply following the time-honored tradition of other disciplines, why all the fuss?

To help explain the uneasiness about the commercialization of genetics, I shall review the scope and nature of the transformation that took place in the biological sciences over a ten-year period between the mid-1970s and mid-1980s. It is certainly not the first time there has been a critical debate over the commercial ties of academic scientists, but this has been the most sustained and highly visible of such debates in this century. Several hypotheses may help to explain this.

• In the post-Watergate, post-Vietnam period, there has been much greater sensitivity and attention given to public and professional standards.

• Biologists have been held to a higher ethical standard and greater public accountability by virtue of the significance their work has for the health professions.

- With biological research financed almost exclusively by public monies, the commercial exploitation of those funds by their recipients became a special target of criticism.
- The entrepreneurial relationships in biology were more aggressive, experimental, and excessive than those in other disciplines, setting new norms for the academic community.
- Media attention to science in general, and the new advances in biology in particular, has intensified; the ties of scientists to the biotechnology industry have been carried on the crest of this attention.
- Federal policies that foster greater linkage between the academic and corporate sectors have attracted public attention.
- The ethical issues associated with the application of recombinant DNA technology have focused special attention on all aspects of the "new biology," including the role of scientists in discovery, development, and evaluation of the technologies.

SCIENCE, ACADEMIA, AND THE VIETNAM WAR

During the Vietnam War, the complicity of universities in weapons research and other forms of military presence on campus such as the Reserve Officers Training Corps (ROTC) were subjects of intense debate both inside and outside of academic circles. Disclosures in the popular left-wing press of classified DOD contracts helped to foment student dissent, which in several cases was expressed in arson and bombings of campus buildings. Any military funding on campus was suspect, although special attention was directed at chemical and biological research. The United States sprayed massive quantities of defoliants that destroyed sensitive ecosystems in Southeast Asia,[3] and dropped napalm [a jelly-like chemical that adheres to the skin and never stops burning] on civilian populations.

The military engaged universities in a wide range of medical, social, and technological research that provided grist for investigative journalism. As an example, a report by the *Village Voice* (October 15, 1971) disclosed that the Pentagon had a contract with a University of Cincinnati medical team to study indigent patients suffering from inoperable cancer who were exposed to lethal doses of radiation. They cited evidence that there was no therapeutic value to the radiation treatment. According to the report—not disputed by the university—the army wanted to learn about reductions in combat effectiveness in troops exposed to lethal levels of ionizing radiation.

A spate of New Left magazines provided periodic exposés of the academic-military-industrial complex. While at the peak of its popularity, *Ramparts* published "War Catalogue of the University of Pennsylvania" (August 1966, pp. 32–40), revealing, among other things, chemical and biological warfare research at the Institute for Cooperative Research.

Despite the keen interest in the university's participation in military research during the Vietnam period, there were no comprehensive studies that examined the types of campus networks that were developed and nourished by the DOD. Universities kept a low profile for these associations, usually to protect their professors and institutional images. The post-Vietnam War period witnessed a sharp decline in DOD funding of academic research, in part a response to changing policies over classified research and military presence on campus. Many colleges ended their ROTC programs during this period.

The university was studied extensively by New Left journalists and social scientists who portrayed academia as an instrument of the capitalist system controlled by the "power elite." Loren Baritz's *The Servants of Power* described how the social sciences were used to support the values of corporate capitalism at the expense of workers.[4] New Left critics argued that the intimacy between the corporate sector and the university resulted in the creation of unofficial taboos about what can be studied. Steve Halliwell, a member of the Students for a Democratic Society, offered this view of the university in his analysis of the Columbia University student rebellion.

To achieve their goal in modern society, the ruling class must have a section of the working population skilled in advanced technology. . . . Technicians are needed, specialists who can develop and administer the systems that will make the society run as smoothly as the massive corporate conglomerates that sit on top of it. The universities have become the training ground.[5]

In contrast, linguist and intellectual historian Noam Chomsky defended the university as a potential but unactualized source of objective scholarship and reasoned analysis of how social change can come about. In his widely cited essay "Knowledge and Power: Intellectuals and the Welfare-Warfare State," Chomsky argued against writing off the university as a fiefdom of the state and a hostage to capitalist ideology.[6]

However, there were stark contradictions between how the university was idealized and what was being revealed about its governance. David Horowitz, analyzing the role of foundations in supporting major university centers, noted:

With few exceptions, of course, these major university research complexes coincide with the strongholds of the old wealth, the aristocratic centers of the American upper class (Harvard, Yale, Stanford, etc.). It is here that the channels to Wall Street and Washington are most open and inviting to the co-optable professor, and the social attitudes and traditions exert the most powerful and most subtle conservatizing pressures.[7]

The idea that universities were colonies of a state-corporate nexus was a persistent theme in the New Left analysis. In the post-Vietnam period however, there was a change in the rhetoric. The concept of colonization lost its appeal. In its place, supporters of an independent university appealed to concepts like influence, governance, and conflict of interest to advance their position.

A STUDY OF SCIENTIFIC ADVISORS TO GOVERNMENT

Most practicing scientists in physics and chemistry are well aware of the multiple roles of some of their colleagues in government and industry. Some

Table 4.1
Academic Scientists with Corporate Affiliations Serving on Federal Advisory Boards, 1957–1973

	Academic Scientists	Class A Connections	Class A or B Connections
President's Science Advisory Committee	55	53%	69%
National Science Board	62	37%	53%

Source: Charles Schwartz, "The Corporate Connection," *Bulletin of the Atomic Scientists* 31 (October 1975): 15.

institutions regard science policy work quite highly. To shape and influence public policy is a sign of prominence and effectiveness. Much of the information in this area has been anecdotal. Charles Schwartz, a Berkeley physicist, made an attempt to quantify what he believed was a disturbing trend within science, namely, government science advisors who did not represent independent expertise. Writing in the *Bulletin of the Atomic Scientists*, Schwartz described a study he undertook that examined the multiple roles of academic scientists as consultants to corporations and advisors to government.[8]

Schwartz applied two criteria in gathering his data. Class A corporate connections described those academic scientists who served on the board of directors of large corporations ($100 million in assets or annual sales). Class B corporate connections described those academic scientists who were on the board of directors of medium to large corporations ($10–100 million) or who were consultants but not board members.

In the first phase of his inquiry, Schwartz examined the multiple affiliations of members of the prestigious President's Science Advisory Committee (PSAC) between 1957 and 1973. In its 16 years the PSAC had 78 members, 55 of whom were identified as primarily academic. The study reported that 53 percent of the academic PSAC members had Class A connections and 69 percent had Class A or Class B connections.

For the second phase of the study, Schwartz looked at the membership of the National Science Board (NSB), which advises the National Science Foundation. He used the same time span (1957–1973) as for the PSAC study. The results indicated that 37 percent of the academics had Class A connections and 53 percent had Class A or Class B connections (see Table 4.1).

From this data and information on other science advisory groups, Schwartz concluded that "many high-level bodies concerned with national science policy have their largest representation drawn from academic ranks . . . however, a large fraction of these academics (generally a majority of them) turn out to have substantial personal connections with large corporations."[9]

This study by a physicist crossing disciplinary ground into social science became obscured very quickly. Nevertheless, it is illuminating in two respects.

First, it focused attention on the myth of the disinterested expert. Second, it raised concerns about public disclosure and conflict of interest regarding scientists at a period of time when the Watergate episode still loomed like a dark cloud over the collective national conscience. Granted the study was small and inconsequential. However, it does provide a backdrop for subsequent studies of academic and corporate linkages in biotechnology. Later in this chapter I describe a similar methodology in a study of multiaffiliated biological scientists who serve on corporate public advisory committees. It is clear from this early investigation that the role of academics in biotechnology followed a preexisting template that was forged by other disciplines.

A decade after the Vietnam War ended, campus sensitivity to the moral status of the university changed. The political and economic climate external to the university brought a new set of incentives that favored aggressive entrepreneurship. The changes took place without much turbulence, though several universities experienced a mild identity crisis, and when they subsided the norms of engagement between campus and board room had been restructured.

THE UNIVERSITY'S MULTIPLE PERSONALITIES

I have conceptualized the American university as an institution with multiple personalities.[10] These personalities coexist in a delicate balance; changes in that balance affect the norms of faculty involvement in external activities. Each of the following personalities represents a form of institutional identity accompanied by its own set of goals and responsibilities. When conflicts arise over university-industry relations, they often reflect more deeply rooted differences among the archetypal personalities.

- *Classical form*: Knowledge is Virtue. In its classical personality the university is viewed as a place where knowledge is sought for its own sake. The direction of inquiry is internally driven and bound by the norms of universal cooperation among scholars.

- *Baconian ideal*: Knowledge is Productivity. The main function of the university under this model is to provide personnel and intellectual resources for economic and industrial development. The pursuit of knowledge is not fully realized unless it can contribute to industrial productivity. The responsibility of the scientist begins with discovery and ends with application.

- *Defense model*: Knowledge is Security. University laboratories and the scientists who manage them are viewed as critical resources for national defense. Campuses that restrict classified or weapons research, or that do not accept ROTC programs and CIA funding, represent barriers to the fulfillment of this model.

- *Public interest model*: Knowledge is Human Welfare. The university exists to solve major societal problems such as dread diseases, environmental pollution, and poverty. Professors are viewed as a public resource and are called upon to tackle complex medical, social, economic, and technological problems.

By distinguishing these institutional personalities, we can draw attention to the contesting value imperatives and responsibilities associated with them. Generally, these personalities coexist in some form of dynamic equilibrium. Factors external to the institution are responsible for a resetting of the balance point. For example, in recent years the classical and public interest identities of the university have given way to the industrial and defense models.

THE STATE'S INFLUENCE ON THE CHARACTER OF THE UNIVERSITY

With the passage of the Morrill Act in 1862 establishing land grant colleges, the federal government began to shape the goals and character of the American university. However, since the start of World War II, federal involvement in both public, land grant, and private colleges and universities has increased dramatically. In the 1940s, scientists at leading universities were mobilized in a secret national effort to develop the atomic bomb. Since that time, the Department of Defense has been a major funder for certain areas of university research, both basic and applied. By the 1980s more than 50 percent of the R&D support for academic disciplines with strategic importance to the military like physics, computer sciences, and electrical engineering came from the DOD.

Figure 4.1 shows the DOD's funding pattern for university R&D over a twenty-year period. In 1968, at the peak of the Vietnam War, the DOD support to universities was nearly $700 million (1982 dollars). The DOD funding fell precipitously in the early 1970s, bottoming out to almost $400 million before it began to rise again. By 1986 the DOD support to campuses was approaching $1 billion.[11]

Figure 4.2 shows all federal funding (including DOD) for basic research (carried out primarily at universities) between 1967 and 1987. Basic research funding (in constant dollars) has risen from about $5.5 billion to $7 billion, with periodic oscillations.

In the area of biotechnology, total federal support for R&D (which includes basic and targeted research) in FY 1987 was approximately $2.72 billion.[12] Most of this funding (83.5 percent) comes from NIH (see Figure 4.3).

Funding in biotechnology represents 38 percent of NIH's total 1987 budget. Within little more than a decade after the critical discoveries were made in applied genetics, biotechnology became a significant focus of federal funding programs. Any field in which molecular genetics could be applied became fashionable.

It is particularly notable that after the NIH, the DOD is the largest federal source of biotechnology funds. In 1987 the DOD funded $118.8 million in biotechnology projects, divided nearly equally between intramural and extramural projects and between basic and applied research. About 85 percent of its extramural research is conducted at universities.[13]

Whether in biotechnology or other fields, the federal government has been the primary source of funding to universities. The rise in U.S. military expend-

Figure 4.1
DOD Funding for University R&D, FY 1968–1987 (1982 Dollars)

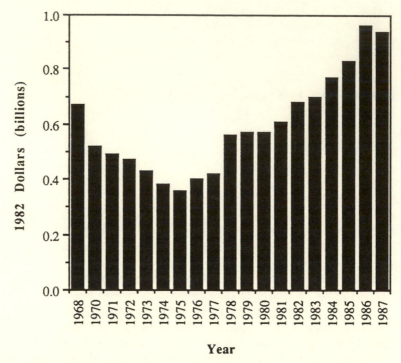

Year

Source: NSF, *Federal Funds for Research and Development* (various editions). Prepared by
    Louisa Koch.

itures for biotechnology may well be within the parameters of the Biological
Weapons Convention, which prohibits signatory nations from producing or stockpiling biological weapons.[14] Nevertheless, some view the DOD's strong interest
in this area as a provocation to other nations, some of whom are not signers to
the convention.

Federal science funding has created a university culture that is symbiotic to
government agencies. The introduction of genetic engineering did not change
that. However, in the early Reagan years, the federal government enacted laws
and established policies that made universities more fertile ground for corporate
investors. In governmental circles the operative phrase was ''university-industry
research partnerships.'' The timing for biotechnology was ideal. This was a
nascent industry building heavily on the intellectual capital of academic scientists.
During this period, a new administration advanced a theory of economic development that was based on stronger cooperation between the industrial and
university sectors. The government saw itself as an important matchmaker in
this courtship of science and technology.

Presidential science adviser George Keyworth, II, explained the declining

Figure 4.2
Federal Funding for Basic Research, FY 1967–1987 (1982 Dollars)

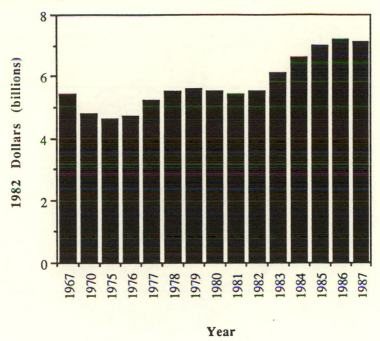

Year

Source: NSF, *Federal Funds for Research and Development* (various editions). Prepared by
   Louisa Koch.

competitive position of the United States by citing this nation's delay or failure
in bringing new scientific discoveries into industrial use. Keyworth exclaimed
that "most academic and federal scientists still operate in virtual isolation from
the expertise of industry and from the experience, and guidance of the marketplace."[15] The separation of academia from industry was cited as a root cause
of the economy's sluggishness.

   Under the Reagan administration, new federal patent legislation passed in
1980 gave universities and small businesses greater incentives to capitalize on
faculty discoveries funded by federal grants. The legislation relaxed criteria for
federal approval of licensing agreements between universities and private businesses. In the same year, a revision in the tax laws created the Research and
Development Limited Partnership (RDLP), a financial mechanism for attracting
R&D investments to university campuses by providing special tax shelters and
high investment income to investors in state-of-the-art technology.

   The Office of Productivity, Technology, and Innovation (OPTI), created in
1981, promoted the use of RDLPs at universities as a means of generating
alternative sources of research support and accelerating the transfer of federally
funded technology. The Economic Recovery Tax Act of 1981 allowed a 25

Figure 4.3
Federal Support for Biotechnology R&D, FY 1987

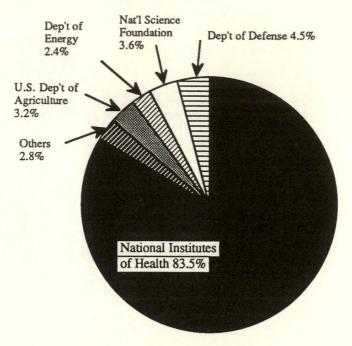

Source: Office of Technology Assessment, *New Developments in Biotechnology: U.S. Investment in Biotechnology—Special Report 4*, OTA–BA–360 (Washington, DC: U.S. Government Printing Office, 1988), 37.

percent tax credit for 65 percent of a firm's payments to universities that support basic research.[16]

A significant percentage of the federal R&D budget goes to the private sector. Some of this funding may result in new discoveries. A persistent question in such arrangements is: Who owns the patent? The Technology Transfer Act of 1986 prompted NIH to establish guidelines that would give companies exclusive licensing rights to discoveries arising from federally funded research and encourage scientists to seek commercial applications for their work. These were some of the leading external incentives that contributed to a new entrepreneurial climate on campuses.

Anticipating tighter research budgets through the 1980s, universities easily accommodated to the new financial arrangements with the private sector. During this period some of the largest financial collaborations were negotiated between universities and corporations in electronics and biotechnology.[17] Prompted by the new relaxed federal rules over patent rights, many of the major research universities began to develop in-house expertise for identifying patentable subject matter on campus. Table 4.2 shows the patents awarded to 22 universities be-

Table 4.2
Genetic Engineering Patents Granted to U.S. Universities, 1973–1984

University	Patents Awarded	University	Patents Awarded
Univ. Calif.	13	Michigan State Univ.	1
Harvard College	9	Florida Atlantic	1
MIT	6	Medical College GA	1
Stanford Univ.	5	Univ. Illinois	1
Univ. Minnesota	3	Univ. Iowa	1
Univ. Texas	2	Univ. Nebraska	1
Columbia Univ.	2	Univ. Rochester	1
Cornell	2	Univ. Toledo	1
New York Univ.	2	Univ. Wisconsin	1
Univ. Michigan	2	Yale Univ.	1
Montana State Univ.	2		

Source: Office of Technology Assessment and Forecast, *Technology Profile Report, Genetic Engineering* (Washington, DC: U.S. Patent and Trademark Office, 1984).

tween 1973 and 1984 in the category of genetic engineering (defined as rDNA technology, hybrid or fused cell technology, protein synthesis, and manipulations of nucleic acids). Universities captured 26 percent of all patents in genetic engineering of U.S. origin awarded during this period. Academia was clearly emerging as a strong player in the race to commercialize genetics.

THE BUSINESS OF SCIENCE ENTERS THE MEDIA

Several key events focused media attention on the growing corporate ties of universities and their faculty in the field of biotechnology. First, there was the announcement in 1980 that Harvard University planned to invest in a commercial biotechnology venture—one in which it would hold a minority interest. A few of Harvard's faculty were to be involved. Second, there were disclosures in the press of professors who sited their private firms on campus or who had difficulty distinguishing federally funded from commercial projects. Third, there were reports of multimillion-dollar contracts and exclusive licensing agreements between multinational corporations and universities. Fourth, there were stories of scientists who became millionaires overnight in pursuit of product-markets for their discoveries.

Because the commercialization of genetics was advancing at such a rapid pace, the frequency of university-business collaborations attracted more attention to them than might have existed had the episodes been spread over a longer time period. Also, the first generation of products that prompted the new partnerships were primarily in the health field—vaccines, scarce proteins, blood factors, and anti-cancer agents. A profit race over discoveries of immense importance for the treatment of dread diseases irritated the public sensibility. Biomedical scientists were held closer to the standard of Albert Schweitzer than Thomas Edison. In the public mind, making a fortune off of an AIDS vaccine is viewed differently than making a similar fortune from transistors. The fact that the new biotechnology was the direct descendent of publicly funded research also fueled the public's skepticism about the pure intentions of science.

In 1979, just before the issue of university-corporate ties drew widespread media attention, the British journal *Nature* brought David Baltimore and myself together in a room, alone except for editors, to engage in a dialogue on the growth of entrepreneurial activities in academia. We spent several hours discussing the significance of university-corporate links in the biological sciences. *Nature* edited the tapes and published a two-page account.[18]

Baltimore is the quintessential multivested scientist. In 1969 he co-authored a document under the aegis of the Union of Concerned Scientists that called attention to the threat of biological weapons. He was awarded the Nobel Prize for his role in the discovery of reverse transcriptase—an enzyme that transforms RNA into DNA. He served on the NIH Recombinant DNA Advisory Committee (RAC). In 1981, while Baltimore was still on the RAC, it was disclosed in the press that he was a company director and the second largest individual shareholder of Collaborative Research, a biotechnology firm located in Waltham, Massachusetts.[19] In the mid-1980s, Baltimore became head of the Whitehead Institute, an independently funded research center in molecular biology with strong links to MIT, while also holding his tenure and professorship in biology.

In our dialogue, I raised warnings about undue corporate influence on biological research, particularly in skewing the research agenda. Baltimore expressed support for the symbiosis of industry and academe. He cited the chemical industry as a paradigm case. But he did raise some caveats about the relationship. Scientists, he said, should remain "entirely open" about industrial affiliations; academics should remain consultants and not turn their laboratories to the service of corporations; students should not be groomed for predefined industrial roles, but allowed to explore their own scientific problems.

The critical difference in our perspectives is that Baltimore believed scientists could do it all; serve their universities, serve industry, serve society, and serve themselves. In his view, if done properly this was an all-win situation. I, on the other hand, contended that the emerging corporatization of biology would result in inevitable conflicts of interest within universities, would skew research, and would produce a generation of scientists whose concerns would shift from public to private interests. Within a year of the publication of the dialogue, the media began reporting allegations of universities selling out, of scientists using their academic positions to promote private interests in biotechnology, and major corporate investments in universities.

The popular press treated the corporatization of biology with skepticism and some cynicism. This was clearly apparent from the headlines of many newspaper, journal, and magazine accounts of the new entrepreneurial activities.[20]

CORPORATE LINKS WORRY SCHOLARS

THE SELLING OF SCIENCE

THE GENE MERCHANTS

BIOTECH: SECRECY AND QUESTIONS IN GOLD-RUSH ATMOSPHERE

DEAFENING SILENCE FROM GENETIC ENGINEERS ABOUT COMMERCIAL
   THREATS

BIOLOGY LOSES HER VIRGINITY

NEW COMPANIES COULD TURN ACADEMICS INTO TYCOONS

BIOMEDICAL RESEARCH ENTERS THE MARKETPLACE

CAN ACADEMIA ADAPT TO BIOTECHNOLOGY'S LURE[21]

With the media publicity mounting, it was predictable that Congress would find some reason to hold hearings. The hook into the issue for Congress was twofold. First, there was the issue of commingling of funds—scientists using publicly funded facilities and researchers for commercial R&D work. Second, there was the concern that industry would appropriate the results of publicly funded research, making the public pay twice for its investment.

Intermittent hearings were held between 1982 and 1985.[22] University administrators testifying before Congress were unanimous on one point: they were perfectly capable of managing their institutional stresses from the new commercial ventures, and government interference represented a threat to innovation and academic freedom. A few universities issued guidelines to faculty on consulting arrangements and entrepreneurship. Most were determined to capitalize on biotechnology funding and not to thwart faculty interest in developing financial partnerships with the new industry. Despite all the investigative news reports and congressional hearings, very little was known about the scope and character of the new biological commercialization and the impacts it was having on academic research and the scientific culture. One federally funded study began to address these questions.

THE HARVARD BIOTECHNOLOGY STUDY

In 1984 a team of researchers headed by David Blumenthal of Harvard University's Center for Health Policy and Management undertook a two-part study of the growth and impacts of university–industry research relationships. The first part of the study involved a survey of biotechnology companies. The investigators sought information about industry profiles, dollar investments in the university, perceived costs and benefits of those investments, and the patents and/or trade secrets derived.

For the second phase of the study, biomedical scientists were surveyed to determine how the commercialization of biology was affecting the nature and climate of their research. Since I shall be reporting on a Tufts University study of university linkages in a later section, for purposes of comparison I will summarize the research design of the Harvard Biotechnology Study.

Blumenthal and his team constructed a list of 435 non-Fortune 500 (NF5) companies that support or conduct biotechnology research, and 115 similar Fortune 500 (F5) companies in four industrial sectors: pharmaceuticals, petroleum products, chemicals, and agriculture. Their criteria for biotechnology research included work in rDNA, monoclonal antibodies, gene synthesis, gene sequencing, cell or tissue culture techniques, fermentation technologies, large-scale purification, or enzymology. From their original list they constructed a random sample of 129 NF5 and 76 F5 firms. After screening for eligibility, they completed interviews with 106 firms (71 NF5 and 35 F5). If we define new biotechnology firms (NBFs) as companies founded after 1972, 71 percent of the NF5 companies fell within this definition.

In the second part of the study, the investigators surveyed over 1,200 faculty in 40 U.S. universities.[23] From their results the investigators estimated that 46 percent of all firms in the biotechnology industry support biotechnology research in universities and that a significant minority of companies is heavily dependent on universities for their biotechnology research. They learned that about one fifth of the companies that are associated with academia contribute 20 percent of their R&D funding to universities. Also, industry's funding of biotechnology research in universities constitutes 20 percent of all available funding in this area. These data confirmed the existence of a strong symbiosis between universities and the new biotechnology industry so aptly symbolized by the aphorism "partners in progress."[24]

Critics of university-industry research relationships (UIRRs) proclaimed that openness of university research would be compromised as a consequence of unfettered industry support. Biochemist Keith Yamamoto of the University of California at San Francisco wrote that "companies formed by individual faculty members . . . have affected adversely both communication and morale within the academic setting."[25]

Until the Harvard study there was only anecdotal information about the effects of university-industry ties (institutional or individual faculty) on openness and the free exchange of information. Blumenthal's team asked companies whether their support of university research resulted in trade secrets. From responses to the query, they estimated that 41 percent of all biotechnology firms involved in UIRRs have acquired at least one trade secret from their university-sponsored research.

The investigators defined trade secrets as proprietary information "protected through systematic attempts to prevent disclosure, including prohibiting publication of research results."[26] Thus, growth of trade secrets is a reasonable proxy measure for erosion of free and open communication. However, from the firm's standpoint, trade secrets are a measure of the commercial benefits of UIRRs.

In the second phase of the study, the investigators surveyed individual faculty on trade secrets and the likelihood that the new arrangements might "impede the free, rapid, and unbiased dissemination of research results." Their results were unambiguous.

Biotechnology faculty with industry support were four times as likely as other biotechnology faculty to report that trade secrets had resulted from their university research. . . . When biotechnology faculty who do not receive industry support were asked whether UIRRs pose the risk of undermining intellectual exchange and cooperation within departments, 68 percent said they did so to some or to a great extent. Among their colleagues with industry support, 44 percent agreed.[27]

The Harvard biotechnology study also found support for the hypothesis that UIRRs skew the research agenda: "faculty members with industry support were more than four times as likely as faculty without industry funds to report that such considerations had influenced their choices to some extent or to a great extent."[28] Nearly 80 percent of the respondents stated that UIRRs pose the risk of placing too much emphasis on applied research.

Another question explored in the study is: How does industry affiliation affect the professional activities of scientists? The survey found that faculty receiving industry support were professionally as active or more active (e.g., publications) as their counterparts without industry support, debunking the view that biting the commercial apple results in a fall from scientific grace.

The Harvard study provides the first important quantitative information about some widely debated conjectures related to university-industry linkage. The information obtained in both the industry and university surveys was based on self-reports. Any bias in reporting would favor a positive attitude toward the linkages since scientists who play a dual role as academics and consultants to industry are likely to underplay concerns about conflicts of interest, trade secrets, and skewing of research. Even if we accept the results as understating the situation, they contribute some essential clues in understanding the transformation that biology has undergone.

The scale of university-industry linkage is also of particular importance if one is interested in the degree to which molecular genetics has been influenced by commercial partnerships. The Harvard study gives us a few clues on the question of scale. Of the 800 scientists interviewed who were doing research in the new biotechnologies, 23 percent indicated that they were principal investigators on grants and contracts from industrial sources. Receiving grants and contracts represents one form of academic-industry affiliation. Other associations include consultant activities, equity holdings, and participation in the development of the firm. Another study, conducted by a group at Tufts University, examined the extent to which life sciences faculty have "formal ties" to the biotechnology industry. In this study our aim was to develop a quantitative measure of faculty-industry networks in North America.

THE TUFTS BIOTECHNOLOGY STUDY

In 1982, historian of technology David Noble and I participated in a symposium at MIT on "The University and Private Enterprise." During my exploratory talk on university-industry ties in the life sciences I fancifully asked the audience to imagine a great map of the United States and Canada showing the location of major universities and biotechnology companies. On this map, lines radiating from the universities to firms would signify scientists with working relationships to companies. This began as a rhetorical device to highlight a moral argument I was investigating. That argument concerned the disappearance of a class of independent academic scientists who could be viewed as approaching the problems of biotechnology without having financial interest in its development.

The more I thought about this hypothetical map, I realized that it could be a valuable source of information for anyone wishing to understand the transformation of biology and the scope of academic participation in the development of the new biotechnology industry. In 1985 I began working on a methodology for constructing a database that would provide an objective indicator, one that does not rely on self-reporting, of faculty participation in the biotechnology industry. Only those university faculty with formal ties to a biotechnology company were to be considered. For the purpose of the study, a faculty member has a formal relationship with a firm if the individual satisfies one or more of the following conditions: 1) serves as a member of a firm's scientific advisory board (SAB) or as a standing consultant; 2) holds a managerial position in the firm; 3) possesses substantial equity in the firm—sufficient equity to be listed on the firm's prospectus; 4) serves on the board of directors of a company.

A composite list of U.S. and Canadian biotechnology firms was constructed from several sources, including trade association lists, published inventories, OTA studies, and media accounts of the industry. Firms were selected for the database according to two criteria: First, they must be involved in the microbiology, genetics, or biochemistry of cells. Companies that specialized exclusively in bioprocessing, fermentation, large-scale purification, or instrumentation were excluded. Second, entries consisted either of newly established (post-1973) companies, their subsidiaries, or large established companies (LECs) that initiated research and development in applied genetics in response to the genetics revolution.

A database compiled between 1985 and 1988 was constructed in two domains. The first domain contained 889 U.S. and Canadian biotechnology companies while the second was comprised of 832 biomedical and agricultural scientists (including plant pathologists, microbiologists, geneticists, and biochemists) with formal relationships to those firms. Of the 889 firms, 286 (32 percent) were public, 406 (46 percent) were private, and 197 (22 percent) had an undetermined status. More than half of the firms (493) were found to be newly established (post-1973) or to have a newly established division or subsidiary.

Public firms are required under federal law to file prospectuses and issue periodic reports to the Securities and Exchange Commission. Lists of scientific advisory boards were obtained from these sources. In contrast, private firms are not required under the law to publish information about their personnel or their financial status. A mail survey was used to obtain information about faculty participation on the scientific advisory boards of private companies.

The data collected over a four-year period gave us a linkage map of dual-affiliated biotechnology scientists (DABS). Although many private firms consider their SAB membership as proprietary information and the status of firms is constantly changing, a data map with nearly 900 dual-affiliated scientists reveals important structural relationships between a burgeoning industry and the academic sector.

Demographic analysis of the biotechnology industry reveals a high concentration of new biotechnology firms or new research divisions of established firms in California, New York, Massachusetts, Maryland, and New Jersey, with nearly 40 percent of the firms located in California (see Figure 2.3). The location data on new firms are consistent with the generally accepted view that the industry established a foothold around major academic institutions.

Financing for biotechnology start-up companies was frequently sought without products, prototypes, or even patents in hand. New firms used their scientific advisory boards as one means of building confidence among venture capital investors in the commercial promise of ideas. University scientists contributed substantially to the promotion and development of new firms. Table 4.3 shows the number of DABS, the total linkages of faculty to the biotechnology industry (since one faculty member may be involved in more than one firm), and the number of affiliated firms for the 24 universities with the largest number of commercially involved faculty in the period 1985–1988. Harvard leads with 69 DABS, 83 linkages, and 43 firms. Following Harvard are Stanford and MIT.

The ratio of DABS to the total biotechnology faculty at an institution provides a measure of industry penetration into academia. However, some institutions with medical schools offer courtesy appointments to clinical practitioners, making it difficult to draw comparisons among institutions. Also, the total number of DABS may be distributed over many departments, some of which can be marginal to the commercial development of biotechnology.

In seeking a better indicator, I chose to measure the ratio of DABS to faculty in commercially active departments. Thus, MIT (which does not have a medical school) has 23 of its 35 DABS in a single department consisting of 74 faculty, yielding a 31.1 percent penetration. Stanford and Harvard, both with medical schools, have respectively four and six commercially active departments with penetration rates of 19.5 and 19.2 percent respectively (see Table 4.4).

The Tufts data not only show that a large percentage of the biotechnology faculty at leading universities are involved in commercialization, but that within a single institution there are many firms represented. Harvard, at one end of the

Table 4.3
Linkages of American Biotechnology Faculty to Firms, 1985–1988

University	#DABS	#Links	#Firms
Harvard	69	83	43
Stanford	40	51	25
MIT	35	50	27
UCLA	26	30	19
UCSF	24	28	14
U. Wisc.	24	24	19
Yale	22	26	21
UC Berkeley	22	24	16
UC San Diego	22	22	11
U. Texas	21	27	22
U. Washington	21	22	18
Johns Hopkins	20	24	16
Cornell	20	20	15
UC Davis	17	17	12
Baylor	17	18	11
U. Minnesota	16	16	12
Columbia	15	18	15
U. Penn.	15	17	11
NYU	14	15	12
Cal. Tech.	12	15	11
U. Colorado	12	15	10
U. Michigan	11	12	12
Tufts	11	12	11
Rockefeller U.	10	12	12

scale, has biotechnology faculty associated with 43 companies. A small percentage of these firms were started by Harvard faculty. At MIT and Stanford, the number of associated firms is 27 and 25 respectively (see Table 4.3).

Some of the prospectuses of firms stipulate proprietary covenants with SAB members. The optimistic view is that the diversity of affiliations at a single institution is a sign the universities are not dominated by a single firm. Nevertheless, the magnitude of firm representation within the university helps us to

Table 4.4
Extent of Commercial Penetration into Select University Departments

	#Key Depts	#Faculty	#DABS	%Penetration
MIT[a]	1	74	23	31.1
Stanford[b]	4	82	16	19.5
Harvard[c]	6	156	30	19.2
UC Davis[d]	2	38	6	15.8
UC San Francisco[e]	1	61	9	14.8
UC Berkeley[f]	5	103	14	13.5
U. Washington[g]	2	79	10	12.7
UCLA[h]	4	115	14	12.2
UC San Diego[i]	1	77	9	11.7
Yale[j]	4	126	14	11.1

[a] Department of Biology.
[b] Medical School—Departments of: Biological Chemistry; Genetics; Microbiology and Immunology; Biological Sciences.
[c] Arts & Sciences: Departments of: Biochemistry and Molecular Biology. Division of Medical Sciences—Departments of: Biological Chemistry and Molecular Pharmacology; Cellular and Developmental Biology; Genetics; Microbiology and Molecular Genetics; Medicine.
[d] Departments of: Plant Pathology; Biochemistry and Biophysics.
[e] Departments of Biochemistry and Biophysics.
[f] Departments of: Biochemistry; Microbiology and Immunology; Plant Biology; Plant Pathology; Molecular Biology and Genetics.
[g] Medical School—Departments of: Biochemistry; Microbiology.
[h] College of Letters and Science—Departments of: Cell and Molecular Biology; Microbiology. School of Medicine—Departments of: Biochemistry; Microbiology.
[i] Department of Biology.
[j] Departments of: Biology; Molecular Biophysics and Biochemistry. Medical School— Departments of: Human Genetics; Cell Biology.

understand the emergence of a new climate in biology where the norm of limited secrecy has replaced free and open communications.

The National Academy of Sciences (NAS) is the nation's preeminent scientific society. Members of the academy provide an important service to Congress and other governmental bodies as expert consultants rendering advice on a wide range of scientific and policy issues. We used the database to obtain a lower bound of NAS members with commercial affiliations. Out of a total of 359 NAS members who can be classified as biotechnology scientists (as of 1988), 132 (37 percent) appear in the DABS database. Since membership in the academy is lifelong, the effective percentage of currently active NAS members with industry affiliation is probably significantly higher than what we found. One NAS member estimated that, for active members, the number of DABS in the NAS is well over 50 percent.[29]

PEER REVIEWERS WITH COMMERCIAL INTERESTS

Peer review is an essential part of the international system of science. It is difficult to imagine the organization of science as we know it without a peer review process. Not only does it help improve the quality of published papers, but it also plays an invaluable role in allocating federal research.

Since commercialization has been so widespread in the biomedical sciences, we can expect this phenomenon to be reflected in the peer review process. The Tufts study examined National Science Foundation (NSF) peer reviewers for 1982 and 1983 and compared this sample with the DABS database. Of those on the DABS list, 49 percent were cited as potential NSF peer reviewers; of the 832 DABS, 343 (41.2 percent) actually reviewed one or more proposals during the two-year period.

It is extremely difficult and perhaps practically impossible to prevent people who are so inclined from pilfering confidential ideas while they serve as peer reviewers. The integrity of the system depends upon informal rules and tacit ethical principles shared by members of the scientific community. But when a significant percentage of the peer reviewers are also involved in the commercial aspects of the field, the incentives to violate the ethical code are greatly enhanced. Some scientists may choose not to publish in the open literature and not to seek grant support if they feel that their ideas may be stolen through the peer review process.[30]

THE IRON TRIANGLE OF BIOMEDICAL SCIENCE

It is no longer possible to draw clear lines of distinction between academic, government, and industry scientists. Federal law permits government scientists to form financially lucrative liaisons with the private sector under the aegis of promoting technology transfer. Universities invest in techniques developed by their own scientists. And professors maintain active academic lives and business relationships in their field of research. The triad of government, industry, and academia constitutes a mutually reinforcing system of self-interest that brings to a close an important period of independence for basic research in the biomedical sciences.

Biologists have indeed matriculated, following their colleagues in chemistry and physics. They have responded en masse to the opportunities of entrepreneurial science. These linkages have appeared rapidly in response to changes in federal law and university policies. Faculty entrepreneurship has also affected the normative structure of science. The concept of free and open communication is no longer the ideal but an impediment in a world where knowledge and investment have been fused. The current environment nourishes conflicts of interest as scientists serve multiple and sometimes competing missions.

Solutions to the problem do not come easily. Several universities have established guidelines. Granting agencies are more cautious about funding investigators who have financial investments in a process or product. And two distinguished journals (*Journal of the American Medical Association* and the *New England Journal of Medicine*) ask contributors to disclose financial associations related to their research. Broader disclosure requirements might reduce abuses or the appearance of conflict. But the greatest loss to society is the disappearance of a critical mass of elite, independent, and commercially unaffiliated scientists to whom we turn for vision and guidance when we are confounded by technological choices. Once the erosion of an independant university sector is accomplished, the stage is set for what University of Washington Professor Philip Bereano aptly described as "the loss of capacity for social criticism."[31]

Part II

Genetics and Ecology

5

Environmental Applications of Biotechnology

Recombinant DNA and cell fusion techniques may soon allow man to change food crops from inside out with great speed and precision, shaping life to fit the environment.

McAuliffe and McAuliffe, 1981[1]

Iowa farmers are divided in their assessments of biotechnology. For the most part, they enthusiastically endorse developments that are expected to improve production efficiencies. But they generally are critical of aggregate increases in food production made possible by biotechnology, as well as likely effects of biotechnology upon farm structure.

Bultena and Lasley, 1990[2]

The concept of diminishing marginal returns applied to industrial or agricultural production implies that increases in efficiency eventually flatten out, at least until a new technological innovation is introduced. The agricultural sector has experienced stepwise gains in efficiency during certain periods in its modern development. The principal innovations have included the electrification and mechanization of the farm, the introduction of antibiotics in animal husbandry, the hybridization of plant species, and the application of chemical fertilizers and insecticides. Most of these innovations have long run their course. With established technologies, increases in efficiency are made at the margins or through economies of scale. Now agriculture is poised for the innovations of biotechnology. Leaders of the major agri-industries are evangelical in their optimism over the possibilities. The president of Monsanto, Earl H. Harbison, Jr., described biotechnology as an "economic engine for future agricultural and industrial growth."[3] Not only has biotechnology been cast as a revolution that will enhance efficiency and growth, and enable the world's food supply to keep pace with the expected increase in population, but unlike its predecessors, it has also been characterized as the revolution that will put humans back in balance with nature. The concept of a ''softer revolution'' has made environmental optimists out of hard-line industrialists. In the words of Monsanto's Harbison: ''We can at last find biological solutions to biological problems that mechanization and chemicals cannot solve.''[4]

The image presented of biotechnology is that the new knowledge of life processes will enable humans to live in a sustainable fashion where economic and ecological efficiency are optimally met. For this to work, the production sector cannot treat nature as a great and infinite wellspring of resources that provides without being provided for.

Have we finally reached a stage in the evolution of technology where new processes of production can be harmonious with natural processes? This expectation carried in the rhetoric of its promoters set biotechnology apart from other industrial revolutions where the criteria for success almost always explicitly favored economic growth and efficiency over the ideas of sustainable production, biocentrism, and ecological balance between humans and the natural habitat. Can biotechnology achieve such goals? Does this revolution embody the seeds of a gentler technology? The four chapters of this part of the book explore the applications of applied genetics to the environment and the social response to several key innovations. Coming as they did in the wake of a new global environmental awareness, the possibilities of releasing genetically engineered organisms into the environment were met with high hopes, skepticism on issues of ecological safety, and some ideological opposition that questioned the wisdom of genetically redesigning plants and animals to increase the human harvest of nature.

THE BACONIAN LEGACY

The desire to control and rearrange species in nature is deeply rooted in the Western scientific tradition. Centuries before genes were understood, scientists contemplated a future in which the natural species were merely starting materials for the conscious rearrangement of the biotic world. For these scientists, the architecture of living forms was likened to the natural minerals of the earth. They were there to be cast into whatever form the human imagination could muster and the life sciences could render.

Francis Bacon, the English scientist-philosopher, unleashed his imagination in a 1622 work titled *The New Atlantis* that describes a biological utopia.[5] Bacon's writings heralded a new philosophy of nature and a new epistemology of science. To have knowledge about the world is tantamount to having power over it. Control over nature was both the goal and the test of knowledge. Bacon provided a literary vision of how scientific knowledge will transform agriculture.

We have also large and various orchards and gardens, wherein we do not so much respect beauty as variety of ground and soil, proper for divers trees and herbs, and some very spacious, where trees and berries are set, whereas we made divers kinds of drinks, besides the vineyards. In these we practice likewise all conclusions of grafting and inoculating, as well of wild trees as fruit-trees, which produceth many effects. And we make by art, in the same orchards and gardens, trees and flowers, to come earlier and later in their seasons, and to come up and bear more speedily than by their natural course they do. We make them also by art greater much than their nature, and their fruit greater and sweeter, and of differing taste, smell, color, and figure from their nature. And many of them we so order, as that they become of medicinal use. We have also means to make divers plants rise by mixtures of earths without seeds, and likewise to make divers new plants, differing from the vulgar, and to make one tree or plant turn into another.[6]

Bacon's "biotopia" included an experimental program for redesigning animals. He described great parks with beasts and birds of all varieties. They were the natural substrate for altering phenotypes to meet human intellectual, practical, or aesthetic needs.

By art likewise, we make them [plants] greater or smaller than their kind is, and contrariwise dwarf them and stay their growth; we make them more fruitful and bearing than their kind is, and contrariwise barren and not generative. Also, we make them differ in colour, shape, activity, many ways. We find means to make commixtures and copulations of divers kinds. . . . We make a number of kinds of serpents, worms, flies, fishes, of putrefaction, whereof some are advanced (in effect) to be perfect creatures, like beasts or birds, and have sexes, and do propagate. Neither do we this by chance, but we know beforehand of what matter and commixture what kind of those creatures will arise.[7]

Bacon's utopian prophecy is remarkable to the extent that it operated on the same paradigm that exists in the scientific community today, namely, that there are no unalterable properties of animals or plants. He rejected the idea of species barriers or the notion of integrated wholes. Furthermore, he recognized no moral considerations in the remaking of life-forms. The new secular science was not burdened with the ecclesiastical sanction that humans should not seek to improve upon God's creatures.

JACQUES LOEB AND THE CREATION OF LIFE

In the latter part of the nineteenth century, German-born physiologist Jacques Loeb advanced the field of biological engineering and believed its fundamental task was to produce living matter artificially. Like Bacon, Loeb refused to accept species boundaries or evolutionary constraints. He actively sought to transcend natural barriers if for nothing more than the challenge of it.

Loeb first received international acclaim for his experiments in circumventing biological fertilization. In 1899, while working with sea urchins, Loeb discovered what he called "artificial parthenogenesis." He devised a method of fertilizing the eggs of sea urchins without male sperm by using inorganic salts. In Philip Pauly's scientific biography of Loeb, he notes that "artificial parthenogenesis was a vindication of Loeb's hopes and a model for science to come in which biologists would consciously work to reconstruct the natural order to make it more rational, efficient, and responsive to the ongoing development of engineering science.''[8]

Loeb demonstrated that what was believed to be a pure biological function—fertilizing an egg—was reducible to physical chemistry. His discovery brought banner headlines such as "Science Nears the Secret of Life," as well as a host of satirical cartoons. Like Bacon, Loeb rejected the Greek view of science as wisdom and understanding of the world as given. In its place he substituted conquest and transformation as the goals of science, "utilizing the forces of nature to bring about new combinations, creating things which have never been created outside of nature's workshop.''[9] For example, Loeb studied the possibilities of redesigning animals that regenerate certain body parts.

I have succeeded in finding animals in which it is possible to produce at desire a head in place of a foot at the aboral end, without injuring the vitality of the animal. I now have animals that have heads on each end of their bodies; I thus have animals with bilateral symmetry that in nature have different oral and aboral poles. This idea is now hovering before me that man himself can act as creator even in living nature, forming it eventually according to his will.[10]

Loeb developed a specialized vocabulary to describe his research program in biological engineering. "Abiogenesis" meant the creation of living things from simple substances, whereas "heteromorphosis" described how a physical process could be used to create biological mutants (two-headed tubularia). Through a series of experiments on lower animals, Loeb sought a "technology of living substance" (*einer Technik der lebenden Wesen*).

SCIENCE AND THE GARDEN OF EDEN

After the late nineteenth century, practical efforts to fulfill Bacon's utopian prophecy were initiated by plant breeders who sought perfection first through the process of breeding, screening, and selection, and subsequently through techniques of cross-fertilization and crop hybridization. Plant breeders have altered the genetic strains of crops and flowers by methods developed over centuries, and hybrids have been disseminated widely in the environment. The genetic architecture of the hybrid plant is modified by human intervention, but not by human design, since there is considerable randomness and uncertainty to the process. Millions of plants may be screened in search for a few favorable genotypes.

Modern genetics goes a step further than its classical counterpart. No longer is the plant breeder constrained by species boundaries. As Bacon had imagined, plant species may be transformed according to some archetype or blueprint. Seeds, once thought to be the basic units of plants, are displaced from their privileged status by the cellular and subcellular units. A similar change has taken place with respect to animal life. The breeding of animals, which can be traced back to the earliest period of domestication, contributes an artificial form of environmental pressure. Evolution is guided by human intention. Thus a sluggish, overweight sow is the desired phenotype for the feedlot, but not, perhaps, for survival in the wild. Industrial and biological evolution became one and the same. The results are animals that are selected for compatibility with human-centered systems—domestic, agricultural, or food processing.

As a method to direct the evolution of animals, breeding is a slow, limited, and ponderous technology in comparison to what may be achieved by the application of molecular genetics to animal reproduction. Since genes can be inserted directly into the fertilized egg, obstacles of species incompatibility are easily overcome. This was dramatized by several experiments on rodents. The first transgenic experiment on an animal embryo took place in 1981 when a rabbit globin gene was inserted into a mouse. In 1982 rat genes and subsequently human genes were implanted into the fertilized eggs of mice. The result was a mouse twice its normal size.[11]

Also, with the methodology of germline manipulation of animals it is possible to create identical clones either by splitting the early embryo or by nucleus transplantation. In the latter case the desired full complement of genes is inserted into a fertilized, denucleated egg. These and other applications of cell mechanics and genetic technology contributed to the commercialization of a new class of environmental products. The following section provides a framework for classifying these research programs in biotechnology as a prelude to an examination of early public policy responses.

GENETICALLY ENGINEERED LIFE-FORMS IN THE ENVIRONMENT

The idea of controlling and improving upon nature has been a persistent theme in the history of technology.[12] It has always been met with its skeptics, who are no less zealous in their opposition to the hubris of science as the proponents of change are to their attack on nature worship. At one end of the continuum are those that argue there is nothing in nature that cannot be improved upon. At the other extreme are those who maintain that evolution, without human intervention, results in the optimality of life processes. "Nature knows best," according to the popular aphorism.

When molecular genetics gave birth to gentechnics, there were already numerous examples of efforts to control and harness nature that had resulted in mixed outcomes. The mining and burning of coal and oil has resulted in new forms of pollution and has been a major contributor to the "greenhouse" effect or the warming of the earth. Synthetic, nonbiodegradable chemicals, which were responsible for the second American industrial revolution, have become ubiquitous residents in the landscape posing remediation problems of enormous magnitude and economic burden. The agricultural system has become hostage to chemical pesticides that continue to pose risks to human health and disrupt natural ecosystems.

The early promise of biotechnology included products and processes designed to ''fix'' the undesirable outcomes of other technologies. It also incorporated research to improve upon existing plants and animals as food sources. In preparation for reviewing the expectations of biotechnology during the first decade, I have classified the environmental applications into five categories: remediation of ecologically harmful technologies; improvement of existing plants; modification of animals; microorganisms to mine the earth; biological sensors.

Remediation

Pollution Cleanup. Research efforts are directed at genetically engineering microorganisms that could degrade oil spills, detoxify chemical pollutants in soil and water, and improve biodegradation in wastewater treatment plants. The problems of off-shore oil spills, toxic wastes at Love Canal, New York, and Woburn, Massachusetts, polychlorinated biphenyl (PCB) contamination of the Hudson River, dioxin in Times Beach, Missouri, and scores of other cases highlighted the importance of dealing with toxic chemicals in the environment. With the tools of classical microbiology, plasmid engineering, and recombinant DNA techniques at their disposal, scientists began experimenting with microorganisms that break down complex hydrocarbons. Early studies showed promise of a bacterium that could digest crude oil. In 1980 General Electric won a patent claim for an oil-degrading strain of *Pseudomonas*, the first such claim issued for a microorganism *sui generis*.[13] Nine years later, the B. F. Goodrich Co. received a patent for a novel plasmid that degrades ethylene dichloride, a compound found in many toxic waste dumps.

Other investigations sought microbes that could neutralize the synthetic herbicide 2,4,5,T (an ingredient in Agent Orange) and PCBs, as well as organisms that could degrade explosive substances and soil contaminated with nuclear materials.[14] The disposal of radioactive wastes has been a persistent problem for the nuclear industry and the Department of Energy. Community resistance to the siting of low-level radioactive waste land fills remains high. The prospect of a biotechnological solution to radionuclides could soften the public's opposition to nuclear power.[15]

Biological Alternatives to Agricultural Chemicals. Promotional materials emanating from the new biotechnology industry spoke enthusiastically of replacing chemical pesticides with biological alternatives that are species-specific and safe for humans and other living things. Some of the new firms embraced the rhetoric of the environmental movement in describing their research agenda and their concern for things ecological. The following is an excerpt from an annual report of the Mycogen Corporation, San Diego, California, circa 1988.

Pesticides are necessary to grow crops economically and preserve the quality of our food supply. Worldwide, a staggering $14 billion are spent annually on crop protection products—$5 billion in the U.S. alone. Despite these huge expenditures, many destructive pests go unchecked and cause major crop losses. The pesticide industry is currently dominated by manufacturers of synthetic chemicals. However, the interplay of environmental and economic forces is creating tremendous pressure for new products which are effective and economical while being safe to use and being compatible with the environment. Biological products are ideally suited to fill the market needs.[16]

Among the major chemical companies, Monsanto was recognized for its strong development effort in agricultural biotechnology.[17] Monsanto's R&D program included the development of a tomato plant genetically engineered with a protein from the naturally occurring bacterium *Bacillus thuringiensis* that is supposed to resist insects. Several environmental groups, including the Environmental Defense Fund, the National Audubon Society, and the National Wildlife Federation supported a project of the firm Crop Genetics that involved injecting a microbe called Cxc into the seedlings of corn. This microbe, which ordinarily lives in corn plants, was endowed with a gene from *Bacillus thuringiensis*, the protein product of which is toxic to certain corn pests.

Another group of agricultural research programs was directed at slowing down the escalating use of artificial chemical fertilizers, which run off into surface waters and foul lakes and streams by the buildup of organic matter—a process known as eutrophication. The expectation was that plants or symbiont microrganisms could be genetically engineered to fix their own nitrogen from the vast reservoir available in the atmosphere. Agracetus began marketing a microbial soybean seed treatment—a product with genetically modified strains of *Bradyrhizobium japonicum* to improve nitrogen-fixing ability. Another strategy involved transferring expressible nitrogen-fixation (nif) genes of organisms like *Klebsiella pneumoniae* into the chloroplasts of plants, endowing the latter with a self-fertilizing capacity.[18]

Agricultural Harvesting and Distribution. Fruits and vegetables are frequently harvested in a preripened stage to accommodate harvesting machines and mass distribution networks. However, the quality of the produce is often affected by being cut off from critical enzymatic changes that take place during the natural ripening process: "While today's tomatoes are the most impressive since the dawn of produce in their resilience, disease-resistance, and travel-hardiness, they are also the most inedible."[19] Some genetic engineering research is directed at modifying crops so that they will be compatible with the harvesting technologies while also retaining the taste and quality of naturally ripened produce. The challenge for biotechnology is to recapture the taste of produce as it was when grower and consumer were in close proximity, and to accomplish this without disturbing the existing agricultural system and by a means that does not compromise the environment.

Improvement of Plants

Adaptability to Harsh Environmental Conditions. Agricultural crops are adversely affected by many environmental factors, including frost, wind, mineral toxicity, soil acidity, salinity, and of course insect pests. A number of research programs were devoted to redesigning plants to be more adaptable to harsh environmental conditions. Kenneth A. Barton and Winston J. Brill saw the possibility of mixing and matching in nature by transferring the stress resistant qualities of weeds to food crops.[20] Among the harvests expected of the technology are disease and drought-resistant strains, self-fertilizing seeds, crops with tolerance to high salinity, and plants with enhanced photosynthetic qualities.

Nutritional Value. Improvements sought for existing crops went beyond their growing characteristics and their resistance to natural enemies. Research was also directed at modifying the protein structure of the crops by altering the amino acid chain in the plant's DNA. Among the expected benefits of this research is the enhancement of the usable protein in staple crops like corn and wheat. In their 1983 article in *Science*, Barton and Brill outlined the possibilities.

The seeds of legumes and cereal grains provide humans directly with approximately 70 percent of their dietary protein requirement. Throughout seed development, storage proteins are synthesized and accumulated within the seed, apparently to provide a source of amino acid reserves during early seed germination. High levels of such protein in seeds provides an enriched amino acid source for both human and animal consumption. However, various deficiencies of seeds in certain essential amino acids do not allow either cereal grains or legumes to provide a balanced diet without supplementation of the limiting amino acids from other sources.[21]

Thus, one approach toward the problem of malnutrition is the improvement of nature's "nutritionally deficient seeds" by encoding them with useful storage proteins. Molecular genetics is viewed as a tool for creating perfect or nearly perfect food sources—foods that carry a complete nutritional balance of amino acids. For lesser developed countries that must import food products to round out the nutritional complement of indigenous crops, or for nations where there has been a scarcity of certain grains, this could be a great asset. But would this technology empower the lesser developed countries? Or would it create a new form of domination by the industrialized nations—a kind of genetic colonialism?[22]

Herbicide Resistance. Nature did not endow most plants with a resistance to synthetic herbicides. Recently, with the extravagant use of herbicides in agriculture and in war as defoliants, plants have been discovered with natural resistance to herbicides. In 1982, herbicides represented 47 percent of the $3.7 billion a year

U.S. pesticide market, eclipsing insecticides and fungicides. Biotechnology offered the herbicide industry a way to expand markets and at the same time create a growing environment that simulated a weed-free green house. By engineering crops with herbicide-resistant traits, the herbicide could be used at any stage in the growing cycle. This so-called improvement upon nature does not take into consideration the effects of the increased use of the herbicide. Guido Ruivenkamp, a Dutch researcher at the University of Amsterdam, wrote that "big industry may see the early opportunity of developing *pesticide-resistant plants* rather than undertaking the longer term effort of developing *pest-resistant plants*."[23]

Plants that Grow Faster. A report issued by the National Academy of Sciences titled *Biotechnology: An Industry Comes of Age* describes an enzyme with the acronym Rubisco that is an important catalyst in photosynthesis. If scientists learn how to move the genes for Rubisco it may be possible to amplify the genes' function and accelerate plant growth. An obvious advantage to the farmer would be the addition of another growth cycle, making more efficient use of agricultural land.[24]

Modification of Animals

Transgenics is the term used to describe the application of rDNA techniques to genetically modify animal species. In some instances, the current research strategy represents an extension of traditional goals to create animals that meet certain requirements of the consumer (as interpreted by agribusiness) or of the food processing industry. For example, there is an interest in creating livestock with leaner meat, and in producing breastier fowl. Since these animals are used exclusively in food production they are not suited to survive anywhere beyond the feedlot. As such they are unlikely to pose environmental problems from a release.

The genetic redesign of animal species has other applications which might affect the environment should the animals be released. There are research programs to use transgenic animals as functioning, living bioreactors. For example, an animal may be genetically engineered to secrete a human protein in its milk or blood stream. In 1987 a transgenic mouse was produced that secreted a clot-dissolving drug in its milk. Mice are produced twice their normal size by the endowment of a human growth gene. In yet another case, the gene that codes for the AIDS virus was implanted in the fertilized egg of a mouse and brought to term. As a result, the AIDS gene is found (albeit not necessarily expressed) in every cell of the mouse. Release of such a mouse into the environment could represent a new route for the spread of the disease.[25] It is for this reason that NIH took special precautions in handling these experimental mice by keeping them in high-containment facilities. There has also been experimentation with transgenic fish to determine if they can be genetically designed to survive in colder water or if a larger species can be introduced to repopulate a river.[26] The lingering question is: what impact would such releases have on the aquatic ecology?

Mining of the Earth

Oil Recovery. Microorganisms can be used to enhance recovery methods for removing crude oil from wells. Initial research suggested that certain organisms can reduce oil viscosity, improve rock porosity and permeability, and decrease water-oil interfacial tension. The technique of Microbiologically Enhanced Oil Recovery (MEOR) was first developed in the 1940s. Through genetic engineering, there was a new impetus to improve MEOR by enhancing the traits of organisms like *Clostridium acetobatylicum*, which has been used in the extraction of oil. The microorganism *Xanthomonas campestris*, which produces xanthum gum, is used for oil recovery by increasing the viscosity of water that is pumped into wells to displace the oil. Another MEOR approach that claims success in recovering oil from marginal wells creates a mixture of microbes and molasses which is injected into oil wells. The gas produced by the microbial growth increases the pressure, forcing out the oil.[27]

Microbial Mining. Microorganisms have also been useful in recovering certain metals from ore-bearing minerals. Copper and uranium are mined and commercially purified by a process that includes leaching by bacteria such as *Thiobacillus ferroxidans*. The leaching process may be done *in situ* or in a controlled environment. Efforts are underway to improve the leaching properties of the organisms through recombinant DNA techniques.

R. H. Zaugg and J. R. Swarz cite some potential hazards in using genetically modified organisms in the mining industry.

All foreseeable applications of biological processes . . . involve microbial systems operating in relatively open environments, such as slag heaps or tailing ponds. Consequently, there are risks that microorganisms or their metabolic products will inadvertently contaminate the local ecology.[28]

The authors offer three examples: (1) bacterial leaching operations that generate large quantities of sulfuric acid could acidify water supplies; (2) *Thiobacilli* and related species may acquire the traits to infect humans; (3) metals concentrated by bacteria from dilute mine waters can accumulate in the food chain.

Biosensors

Microorganisms may also be used as markers in the environment. The concerns about releasing novel organisms have produced an interest in the methods of monitoring the dispersal of the agents. Monsanto Company has developed a genetically engineered strain of *Pseudomonas* carrying natural resistance to antibiotics and genes from *E. coli* that code for proteins that enable the microbe to metabolize lactose. These traits make the organism a possible marker for field tests. The first approval by the EPA of a field test of a genetically engineered organism under the Toxic Substances Control Act (TSCA) was for Monsanto's genetically marked microbe. The test was performed in Blacksville, South Carolina.

The environmental uses of applied genetics constitute an important part of the industrial agenda. But the prospects of releasing genetically novel plants, microbes, and animals have activated a lively public policy debate. The next chapter traces the nature and origins of these concerns and shows the regulatory transition that took place from laboratory experiments to field experiments. The next generation of environmental experiments and applications of biotechnics brought a new group of scientist-stakeholders into the discourse of risks and benefits of new technologies.

6

Evolving Policy: From the Laboratory to the Field

The timely development and rational introduction of R-DNA modified organisms into the environment depend on the formulation of sound regulatory policy that stimulates innovation without compromising good environmental management.

National Academy of Sciences, 1987[1]

Ecologists are unable to predict which introduced species will become established and which will not, and it is often not possible to explain successes or failure after the fact.

Martin Alexander, 1985[2]

The weighty publicity over biotech products and research directions helped bring about a new configuration of public advocacy. Transgenic animals was an issue that linked animal rights organizations, environmentalists, and alternative agriculture groups. The development of more refined genetic screening techniques brought warnings from civil liberties and disability rights advocates. Disclosures of rising expenditures in the Department of Defense's biological defense program stimulated interest from activists in the disarmament community. Feminist–health advocacy groups began exploring the impacts of genetic techniques on reproductive technologies. Bioethicists, clinicians, and religious leaders began tackling the thorny problems of human gene therapy. The prospect of major pharmaceutical advances through rDNA research provided the grist for debates in the international health community on the priorities for developing vaccines. Food and agricultural organizations questioned the impact of biotechnology and the new patent provisions on control over plant genetic resources.

It appeared that every major industrial innovation in applied genetics tapped a wellspring of new issues that were brought to the social agenda. Of these, the issue that ignited the strongest public reaction during the early stages of the

biotechnology revolution was the introduction of genetically engineered organisms (GEOs) into the environment. This chapter examines the origins of regulatory oversight over deliberate releases and the federal efforts at creating an
orderly transition from small-scale laboratory applications of gene-splicing to
large-scale releases of GEOs into the environment.

SOCIETAL CONCERN OVER GEOs

Why has there been so much concern over the risks of releasing plants and
microorganisms that have been modified by genetic engineering techniques?
Shouldn't the emphasis be on the product and not the particular way the product
is created? There are three plausible explanations for the cultural selection of
genetically engineered products as a special area of concern. First, the perception
of risk associated with deliberate release has largely been formed from prior
concerns about recombinant DNA research. In other words, the environmental
problems of genetically modified organisms were inherited from earlier stages
of the genetics debates. Recombinant DNA-produced organisms are what R. E.
Kasperson calls a "social amplifier" in the public's perception of risk.[3]

A second explanation is based upon the notion that genetic engineering provides far greater specificity and control over the product than one could achieve
by plant hybridization or breeding of animals. As a result of the specificity of
rDNA techniques and their capability of joining quite distant forms of life, the
novel life-forms span wider species boundaries and are subject to fewer natural
constraints.

By inserting a single foreign gene, a phenotypic property of a bacterium may
be radically altered. Resistance to antibiotics is such a property. The ability to
change phenotypes in the laboratory with such ease has heightened concerns
about deliberate release. Will nature have an opportunity to accommodate to
these sudden changes?

To the contrary, some scientists have argued that the specificity of genetic
modification makes modern gene-splicing safer than conventional genetics,
where genes get mixed randomly and in large clusters. Following this line of
reasoning, the precision of gene-splicing means that the resulting properties of
the organism will be easier to predict. According to plant geneticist Winston
Brill writing in *Science*, in conventional breeding it is impossible to predict the
properties of the progeny from most of the crosses. Genetically engineered plants
have greater specificity: "If we compare plants derived from breeding programs
with those derived through genetic engineering, it is clear that, in the latter case,
the addition of a few characterized genes to the plant results in properties that
are relatively easy to predict."[4]

From another perspective, despite the specificity of rDNA techniques, with
them one might be capable of producing more substantial changes in organisms
with fewer genetic alterations than with classical genetic techniques. The issue
of whether modern genetic engineering techniques are capable of producing

varieties of plants, microbes, and animals that could never have arisen from the natural rearrangement of genes remains unresolved. It is widely acknowledged that we humans can at least accelerate or redirect the evolutionary process even if we cannot create qualitatively new life forms.

A third reason why genetically engineered plants and organisms designed for environmental release have attracted more concern than the release of similar products prepared through conventional genetics is related to the reputed power of the new technology. Gene-splicing has been the raison d'être of a technological revolution. This is not simply another discovery in the slow, incremental growth of science. This discovery has given birth to a new industrial process for radically reconfiguring biological matter. The disclosure that there is a new power to transform nature is one of the sources of public and scientific anxiety. It might be argued that if rDNA technology embodies a power to stimulate the growth of a multibillion dollar industry, why should its risks be considered comparable to those of conventional genetics? What is the likelihood that the industrial potential of gene-splicing (gentechnics), which is, let us say, a thousand times greater than that of conventional biotechnology, will be unleashed without any increases in environmental risks? It is certainly a question worth considering.

Setting aside for the moment the body of scientific argument about the potential risks of new biotechnologies, there is an undisputed equation between technological power and risk anxiety that must be considered in fully understanding the public reaction to biotechnology. The simultaneous pronouncements about power and safety seem incongruous to a popular culture that has been sensitized to technological failure.

In trying to comprehend the risks of releasing genetically altered species into the environment, inevitably we are drawn to comparisons. Two technologies of commensurate transforming quality to gene-splicing are synthetic organic chemistry and nuclear physics. Both of these technologies are capable of creating new arrangements of matter in a fashion analogous to the creation of novel species through biogenetic engineering.

Risk assessment for synthetic chemicals has been in progress for several decades. There have been some important breakthroughs as well as notable impediments. The identification of a chemical substance is a well-defined process. It is, therefore, possible to construct a precise inventory of chemical compounds. The same is not true for biological agents, at least in some practical sense. Microorganisms and plants are classified by phenotype, and therefore the addition or deletion of a few genes will not necessarily warrant a change in the classification.

It is estimated that 60 to 80 thousand distinct chemicals are used in industry out of a pool of several million that have been synthesized. If a genetic identification system were used for biological organisms, the number of extant chemicals would pale against the number of distinct life-forms since, for the latter, a single nucleotide change would be a differentiating factor. Keeping track of novel organisms and establishing an identification system is a problem of enormous complexity, and probably unrealistic since genetic mutation is a constant occurrence. Yet any serious regulatory effort in biotechnology must address the identity question.

An obvious difference between inert chemicals and life-forms is that the latter are self-reproducing. Throughout the history of the chemical industry there have been countless cases where toxic carcinogenic chemicals were disposed of in lagoons and landfills. These chemicals saturate the soil and eventually migrate to subsurface water supplies where they contaminate drinking water. Once embedded in the earth, many industrial chemicals are difficult to remove. Entire neighborhoods in areas such as Times Beach in Missouri and Love Canal in New York have been evacuated because of toxic waste. In other, more manageable situations, contaminated soil is removed or filtration methods are applied to poisoned wells.

The mistaken release of a nuisance biological agent cannot be handled by techniques developed for chemical contamination. At the worst, the released organisms are beyond recall and will grow to population orders of a magnitude beyond the density of the inoculation. Moreover, if a novel organism were introduced and subsequently found to be dangerous, geographical isolation and community evacuation would simply not work.

Considerable progress has been made in standardizing toxicological testing for chemicals. The use of accepted methodologies, standardized target species such as germ-free mice or rats, and microbial assays such as the Ames Test have contributed to uniform standards of risk analysis. Notwithstanding the progress, there are still many areas of uncertainty and scientific debate. Among them are questions about dosage and extrapolation from animal to human effects. Also, human epidemiological studies are frequently too insensitive to pick up small increases in cancer incidence over a lifetime exposure. While there are many effects of chemical exposure that are not well understood, there is at least a basic methodology for gathering the data.

There is no commensurate methodology for assessing the risks of released organisms. Moreover, the risks associated with certain chemical releases are real. Their effects on humans and the biotic world have been observed. In contrast, the potential risks of genetically altered life-forms are currently speculative. As a result, the social demand for evaluating the risks of bioengineered products designed for environmental use may not evoke the same urgency as if the hazards were confirmed.

When chemicals enter the environment, it is not always obvious what effects the breakdown products (metabolites) will have on the ecology. Chlorine has been used extensively to purify drinking water. Its use has been associated with the appearance of chloroform, a potent carcinogen, and other troublesome compounds called trihalomethanes. While there may be risks in the continued use of chlorine, the alternatives are not good. No safer method for purifying water is available. We can never be sure how released chemicals will reconfigure themselves in ecosystems, but compared to the possibilities of biological entities, the range of unexpected outcomes for inert chemicals is probably much narrower since the biological entities mutate. A subclass of all possible mutations affects the phenotype of the entity. Whether it is more complicated to predict the mutational possibilities of a novel organism than to predict the synthetic pathways and metabolites of a new chemical is the subject of debate. Organisms are certainly more complex than inert chemical compounds. That does not mean that they are inherently riskier, but it does portend a high level of complexity in analyzing environmental risks of genetically novel entities in contrast to newly synthesized chemicals.

There are also some comparisons to be made between biotechnology and nuclear technology. The number of radionuclides is relatively small, surely in comparison to industrial chemicals. They are generally used in well-defined and controlled settings. There are laws regulating the release of radioactive materials. Also, the health effects of radionuclides in high and moderate doses has been studied and is reasonably well understood. This is not the case, however, for low-dose, long-term exposure.

Radioactive materials are detectible in minute quantities with a sensitive monitoring device. To improve the confidence of residents living in the vicinity of a nuclear power plant, some communities have been provided with radiation detection counters.

There is nothing analogous in biotechnology. In theory, one can identify and track bacteria that have been released into the environment. The organisms would have to be tagged in a special way. Even then the identification can be a difficult task depending upon the behavior of the microbes and the conditions of the environment. Each case is unique. At this time, there are no standard methods of detection and no cannonical procedures for distinguishing safe from deleterious organisms. In comparison to bacteria, it is much simpler to detect the spread of an unwanted plant (a weed). But once released, plants, like bacteria, may be impossible to recall.

We have seen how, within a period of a decade, a single critical discovery was the progenitor of an industrial revolution. The investment, scientific, and corporate communities moved expeditiously to capitalize on the commercial opportunities of the new genetics. To whom was this new industry accountable? How were the public policy issues handled? What social guidance was imposed upon the new technological direction? The next section discusses the early regulatory response to environmental applications of biotechnology.

NIH'S EARLY ROLE

Beginning in the 1980s, industry and university proposals for field-testing genetically modified plants and organisms triggered a major science policy debate in the United States that spilled over to the European community. Those who have followed the recombinant DNA controversy from its inception will recognize that the current configuration of policy alternatives is the result of a ten-year historical process. Initially, molecular geneticists cast the problem of genetic engineering in technological terms. Gradually, public perception of the problems associated with gene-splicing focused attention on the ethical and ecological issues. The emergence of a second generation of genetics policy debates brought participation from new disciplines, new communities, new public interest groups, and new federal agencies. Public concerns slowly shifted from the singular issue of laboratory safety to a much broader range of problems. And while these changes were taking place, regulatory oversight of biotechnology also shifted from the National Institutes of Health to other governmental bodies.

For nearly a decade, the agency that assumed primary responsibility for the safe uses of genetic engineering was the National Institutes of Health (NIH). Essentially a science-funding agency under the Department of Health and Human Services, the NIH established the Recombinant DNA Advisory Committee (RAC) in 1974 at the recommendation of Paul Berg, the Stanford University biologist who provided leadership in the early efforts to assess the risks of rDNA research.

According to Berg's plan, an international meeting of biologists (held at the Asilomar Center in Pacific Grove, California, in 1975) would result in a set of principles for safe handling of rDNA molecules. Those principles were then to be used by the NIH's newly formed scientific advisory committee to establish guidelines for all genetic experiments involving the cutting and splicing of foreign genetic material. Berg, along with other scientists who organized the Asilomar meeting, surmised that if the NIH did not act with dispatch in responding to the potential risks of gene-splicing, Congress might pass restrictive legislation. They viewed the passage of such legislation as detrimental to the interests of biology; particularly at stake was the legitimacy and progress of molecular genetics. First, biologists could lose influence over the risk assessment and risk management process. Second, the field of molecular genetics might become stigmatized as the only discipline whose principal research method was regulated. Third, there was considerable concern that rDNA legislation would be inflexible and difficult to amend. Asilomar organizers feared that biology would be saddled with irrational requirements.

Written by scientists for the general use of scientists, the NIH guidelines made their debut in June 1976. No explicit references were made to industrial processes or non-NIH supported uses of rDNA. The guiding principle behind the development of the guidelines was containment. Since one could not predict at the outset with any reasonable degree of certitude that a foreign gene introduced into an organism could not inadvertently transform it into an epidemic pathogen, the consensus at Asilomar was to construct a set of containment provisions consisting of physical barriers, safety operations, and carefully selected host organisms chosen because they do not survive well outside of the laboratory setting. Each containment level was matched with a class of experiments that was permissible under the stipulated conditions. As scientists on the RAC and elsewhere became more confident in the safety of rDNA techniques, laboratory containment requirements were substantially relaxed.

The early NIH guidelines contained a provision that restricted industrial applications of rDNA technology. Cultures of rDNA organisms produced or handled in volumes greater than ten liters (classified as large-scale) were prohibited for NIH grant recipients. The large-scale prohibition was not based upon a scientific assessment of risk. It was a convenient threshold introduced by academic scientists to protect the use of standard laboratory beakers in basic research.

A second provision of the NIH guidelines restricted industrial activity by explicitly proscribing the intentional release of an rDNA organism into the environment. Since absolute containment was nothing more than an idealization, unintentional releases were considered unavoidable. However, by limiting the volume of rDNA culture, the probability of escape could be minimized. Also, since the volume of agent released is correlated to survival and propagation, the large-scale prohibition also supported the general containment strategy.

Commercial interest in rDNA techniques grew rapidly in the mid-1970s.[5] Estimates of industry growth vary. My own count indicates that a minimum of 14 new biotechnology enterprises (NBEs) were formed in 1976, the year the NIH guidelines were introduced. During 1979, NBEs grew by at least 26; and in 1981, at least 66 NBEs entered the biotechnology industry (see Chapter 2).[6]

The NIH had no legal jurisdiction over the research in private companies. In practice, however, firms were not willing to risk the negative publicity that might arise if they violated the NIH guidelines. The nascent biotechnology industry was comprised predominantly of small firms started by relatively young scientists, many of whom retained their university affiliation. The close link between academe and industry may help explain the high degree of compliance among new firms with the standards adopted by the NIH. Since these scientists were groomed on the NIH guidelines, the complications of compliance that might have beset a new industry were minimized. Despite a watchful media, there is no evidence that the NIH guidelines were flaunted by the biotechnology industry. Ironically, the few cases where violations of the guidelines were reported took place at universities. Geoffrey Karny cited two factors responsible for industry's compliance with the voluntary guidelines: "First, the possibility of tort lawsuits provides monetary inducement to comply with the *Guidelines*, which would probably be accepted as the standard of care against which alleged negligence would be evaluated. Second, the threat of statutory regulations, which the companies have sought to avoid, always exists."[7]

Between 1976 and 1979 the NIH process for overseeing rDNA research was put to its severest test. First, the city of Cambridge, Massachusetts, issued a moratorium on rDNA experiments requiring moderately high physical and biological containment. After a widely publicized citizen review process, the city passed the country's first rDNA law in 1977. The law departed from the NIH guidelines in a few minor respects. More importantly, it symbolized the right of local government to exercise control over where the research gets sited and the safety conditions of its performance. Moreover, it established uniform and legislatively mandated guidelines for both publicly and privately funded research.

After the Cambridge rDNA law was passed, nearly two dozen states and local communities debated the issues. Legislation was enacted in about half the jurisdictions. In response to local events and a national mood of concern toward gene-splicing, fifteen distinct bills were filed in Congress between 1977 and 1978 to regulate rDNA research.[8] Some of these bills would have shifted the regulatory authority from NIH to a national commission. These bills also varied in the degree to which local laws were subject to federal preemption. Congress spent two years debating the issue of an rDNA law. A compromise bill was finally voted out of committee early in 1978, but for lack of strong congressional leadership and interest it failed to reach the House or Senate floor for a vote.[9]

Congressional failure to enact legislation strengthened the NIH's position as the sole agency overseeing rDNA activities. Responding to continuing public concern over the research, the Department of Health, Education, and Welfare (HEW, currently the Department of Health and Human Services) rewrote the RAC's charter and increased the size of the committee from sixteen to twenty-five. The new charter, issued in 1978, stipulated that one third of the committee was to consist of individuals with expertise and interest in public health and the environment. The change in the composition of the RAC drew sharp criticism from prominent scientists who argued that rationality was being compromised by including nonscientists in what was essentially a technical process. When the expanded RAC met in early 1979, its agenda was filled with petitions for relaxing containment requirements and approving additional host-vector systems.

Over the next few years several important changes were made in the rDNA guidelines that established a role for the NIH in the review of field tests. A voluntary compliance program was established that gave the private sector access to the NIH review process, while prohibitions against large-scale rDNA activities and the intentional release of genetically altered strains into the environment were removed.

VOLUNTARY COMPLIANCE PROGRAM

After the decision by Congress not to enact rDNA legislation, there was a strong residue of criticism that the NIH guidelines could not protect society from the potential adverse consequences of commercial gene-splicing. In response, NIH developed a voluntary compliance program for institutions that did not fall under the agency's purview. This initiative gave the fledgling biotechnology industry the opportunity to gain the imprimatur from the NIH for both laboratory and commercial-scale rDNA work. A firm wishing to participate in the program first submitted the composition of its institutional biosafety committee (IBC) to the NIH's Office of Recombinant DNA Activities for approval. Once IBC approval came, a firm could file requests with the RAC following procedures similar to those of university petitioners.

One difference between the NIH's handling of academic and industry proposals is that, on the occasion of the latter, the RAC went into closed session. Members were required to sign confidentiality pledges for the protection of information deemed proprietary by the firm. Some RAC members were opposed to having the committee review proposals in closed session. One public interest member refused to participate in the review of industry proposals. He argued that oversight of the private sector was the responsibility of those agencies of government with statutory authority to protect workers, public health, and the environment. Since the NIH lacked authority to carry out these functions, he felt his participation would give legitimacy to this extension of the NIH's role. Another RAC member expressed the following sentiment: "Voluntary compliance is the worst of all possible worlds. . . . You achieve none of the objectives of regulation and none of the benefits of being unregulated. All you're saying is 'I give a stamp of approval to what I see before me without any authority to do anything.' "[10]

The program was also the target of mainstream critiques. As an example, in its 1981 report on biotechnology, the Office of Technology Assessment wrote: "The most significant limitation in the scope of the Guidelines is their non-applicability to industrial research or production on other than a voluntary basis. This lack of legal authority raises concerns not only about compliance but also about NIH's ability to implement a voluntary program effectively."[11]

In May 1979, the NIH's advisory committee went on record opposing voluntary compliance by a vote of nine to six, with six abstentions. The RAC voted that non-NIH funded institutions should be required to comply with the guidelines. This recommendation notwithstanding, the voluntary compliance program became a permanent part of the guidelines in January 1980.[12]

Many biotechnology companies considered it in their long-term interest to secure RAC approval for their rDNA activities. Regardless of how the firms' management felt about the potential risks of gene-splicing, submitting experiments for NIH approval was excellent public relations. The voluntary compliance program helped the biotechnology industry respond to the criticism that the private sector was operating without regulations. When pressed by local communities to demonstrate the safety of genetic experiments, a company's most compelling response was that it complied with the NIH guidelines.

In the wake of sporadic episodes of local opposition to genetic engineering research, the biotechnology industry sought a predictable and stable regulatory climate, but one that would not impede research and development. The NIH contributed to this goal, but with limited success. The voluntary compliance program proved functional to the industry during its early years of development when there was intense competition, investment instability, and the uncertainties of local regulation. As the commercial activities progressed from laboratory research to product development, the NIH policies accommodated to the new stage of industrial activity, particularly in their response to large-scale work and the release of genetically modified plants and organisms into the environment.

LARGE-SCALE PROHIBITIONS

Provisions were built into the early NIH guidelines for waiving the large-scale prohibition in cases where minimal risks were balanced against important societal benefits. The wording of the prohibition was clarified in the 1978 version of the NIH guidelines. No exceptions to the prohibition against the production of large-scale cultures were permitted "unless the recombinant DNAs are rigorously characterized and the absence of harmful sequences established."[13] Under NIH's leadership the risk assessment paradigm was in large measure still under the primary influence of geneticists.

For a short period of time, the NIH gave serious attention to all phases of large-scale work with rDNA molecules. Proposals submitted to the RAC were required to include a description of laboratory practices, specifications on physical and biological containment, risk data, characterization of genes, and the design of the fermentation equipment in conjunction with the physical description of the facility. In 1980, the RAC published a standard that it planned to use for reviewing large-scale proposals.[14]

The committee's role in evaluating plant design and operations for large-scale fermentation drew criticism from some members. They argued that the RAC should confine itself to advising the NIH on the nature of biological procedures and not plant operations. The committee, after all, had no special expertise in bioprocess safety engineering. It depended on outside consultants for guidance in this area. Also, it was a matter of some significance that molecular biologists on the RAC found the review of plant design boring and uninformative. Other RAC members contended that if the voluntary compliance program was to mean anything, a comprehensive review of large-scale operations was essential. Since no other federal agency was engaging in this review, they believed that it was the NIH's responsibility to fill the regulatory void.

The internal debates over this issue were a poignant reminder of the NIH's ambivalence in serving as overseer of commercial genetic engineering. These debates also cast doubt on the logic of having a biomedical funding agency guide a nascent industry exclusively through a system of voluntary measures. The peculiar nature of this regulatory program began to reveal its contradictions. For example, within the RAC opposing factions interchanged positions. Initially, the group most skeptical about the safety of genetic engineering expressed opposition to the RAC's review of industrial proposals, particularly large-scale ones. They reasoned that the NIH was acting as de facto regulator without enforcement powers or congressionally derived authority. If the RAC refused to serve this function, members of this faction believed that agencies like the Occupational Safety and Health Administration (OSHA) and the Environmental Protection Agency (EPA) would enter the field.

Concurrently, there were other members of the RAC who opposed broader federal involvement in biotechnology, particularly legislative or rule-making actions. They viewed NIH's continued role in overseeing commercial rDNA work as essential to preempting such initiatives.

The unexpected reversal took place among committee members around 1980. The pro-regulatory group grew less sanguine about the involvement of other agencies in regulating biotechnology. As a consequence, this faction began supporting a stronger role for the RAC. Meanwhile, the committee's regulatory minimalists, also confident that broader agency involvement was unlikely, backed an NIH role over commercial rDNA activities that was limited to issues of pure genetics, namely, sequence characterization and approval of host-vector systems.

The regulatory minimalists succeeded in so limiting the RAC's review of the safety of rDNA products. By 1981, firms taking part in the NIH's voluntary compliance program were no longer obligated to submit data on sterilization of their fermentation system or their procedures for disposal of rDNA cultures. This policy change was expressed in a proposition adopted at the September 1980 RAC meeting: ''The RAC will determine if a given recombinant DNA-containing strain is rigorously characterized and the absence of harmful sequences is established. Such a determination shall include specification of a containment level. These determinations should not in any way be construed as RAC certi-fication of safe laboratory procedures for industrial scale-up.''[15]

The potential release of rDNA organisms through the effluent of bioreactors did not attract any public attention. While the EPA had jurisdiction over such biological releases under the Clean Water Acts, a constituency for agency action did not develop. The newly formed Committee (later Council) for Responsible Genetics published an article on biogenetic waste in its public interest bulletin *GeneWatch*.[16] However, the issue of biogenetic waste was not a rallying point among environmentalists or the general public. It seemed to lack some important features that influence risk selection. First, it did not impart a clear and present environmental danger. Second, the waste stream, even if it carried genetically engineered organisms, did not excite the media. Bioeffluent was not intentionally designed to alter nature. Since science writers are captivated by the frontiers of science, there was not much of a story in fermentation sludge.

Quite a different media and public reaction took place when proposals appeared before the RAC requesting approval to field-test genetically modified plants and bacteria. There is an informative distinction to be made in the public reaction between intentional and unintentional releases of genetically modified life-forms. To fully grasp the distinction, we have to look at the kinds of experiments proposed, the types of media coverage they were given, the role of scientists in raising safety concerns, and the development of a public interest constituency.

DELIBERATE RELEASE OF GEMs

As the RAC reduced its oversight over industrial bioprocesses, it became more active in reviewing environmental releases of genetically engineered microorganisms (GEMs). Consequently, field-testing was given greater visibility in the media. From 1980 to 1984, when companies and university scientists were preparing to field-test their products, the NIH was the only federal agency with active responsibility over deliberate release experiments involving genetically altered plants or microorganisms.

Progressive relaxation of the rDNA rules finally led to the removal of barriers to field-testing. The first revisions of the guidelines in 1978 still prohibited "deliberate release into the environment of any organism containing recombinant DNA," but individual waivers were permitted after proper public notification, RAC review, and approval by the director of the NIH. The revised NIH guidelines of 1982 eliminated the entire list of proscribed experiments. By June 1983, the prohibition against intentional release of rDNA organisms was replaced by a multitiered review process. Submissions for deliberate release required approval by the RAC, the institution's biosafety committee (IBC) and the NIH director, in addition to various subcommittees.[17]

Agricultural applications of biotechnology, widely publicized in the media, were being readied for field trials. By December 1983, the RAC had reviewed and approved three proposals for releasing genetically altered life-forms. In each case, the committee concluded that the tests posed no significant risk to health or the environment.[18]

The first three proposals for deliberate release came from university scientists and therefore did not involve proprietary information. In March 1980, a Stanford University scientist requested approval from the RAC to field-test a corn plant into which had been inserted the corn storage protein gene with modified sequences. The genetically altered strain included a new corn protein gene that encoded all the amino acids essential to humans (including an enhancement of lysine and methionine, in which corn is deficient). By genetically engineering the new corn protein with the full complement of the essential amino acids, its value to the human diet would be improved. The RAC failed to come up with a hazard scenario for the genetically modified corn; nevertheless, the investigators were required to detassel the plant to prevent pollen dispersion during the field trials. Permission for the test was granted in August 1981. There was no significant public reaction to the decision.

A second proposal for field-testing was brought to the RAC by a Cornell University scientist in June 1982. Tomato and tobacco plants were transformed with DNA from yeast and *E. coli* to provide them with antibiotic resistance. In reviewing the experiment, the RAC raised concerns about the possible spread of antibiotic resistance and the effect of an itinerant plant on the ecology. Neither of these concerns delayed the committee's approval, which it gave on October 25, 1982. When the recommendation reached the director of the NIH, he referred the proposal to the U.S. Department of Agriculture (USDA) for a reading. After USDA approval was given in February 1983, NIH accepted the field test in April 1983, nearly a year after the proposal was first brought to the RAC.

The third of the early field test proposals for genetically modified life-forms proved to be the most controversial. Ironically, it was viewed by some experts as among the safest deliberate release experiments one could perform in the environment. There was one important distinction between the first two proposals and the third: the latter consisted of genetically altering a soil bacterium. The difference between plants and bacteria proved to be a critical factor in the public perception of risk. Also, there was an established tradition of introducing hybridized plants into the environment. However, no analogous tradition existed for microorganisms. The relative novelty of this field test was reflected in the RAC's review.

In September 1982, scientists at the University of California at Berkeley proposed to field-test two soil organisms (*Pseudomonas syringae* and *Erwinia herbicola*) from which about a thousand base pairs of DNA sequences had been deleted. In their natural state these organisms synthesize a protein that provides a nucleation point for ice crystallization, and are known as ice-nucleating agents (INA). By excising the genes responsible for ice nucleation and establishing the genetically modified organism in the environment, scientists believed they could reduce the temperature at which frost begins to form on crops.

When the proposal for field-testing the microbes with the deleted ice-crystallization gene (denoted INA$^-$ or ice minus) was first brought before the RAC, one member expressed concern about the role the organisms might play in the environment. Questions raised about the effect of INA$^-$ on precipitation patterns were based upon some theories that INA$^+$ (normal ice-nucleating agents) played a role in atmospheric weather. The precipitation issue was not resolved by the committee. Nevertheless, the RAC approved the field test, although by a small plurality (seven yes; five no; two abstentions). When the recommendation was transmitted to the director of the NIH, more data along with additional safeguards were requested. Several months later on the second round of review, the RAC approved the proposal unanimously. The NIH gave the final go-ahead for the field test in June 1983, some nine months after it had initially received the proposal.

Investigators agreed to mark the strains of the ice minus organism with antibiotic resistance, which allowed them to monitor the dispersal of the organism. They also agreed to limit field-testing to a single location, the University of California agricultural field station at Tulelake in northern California, a site isolated from the major fruit tree and citrus-growing regions of the state. Despite these precautions, the RAC's decision on the field test for ice minus was met with citizen opposition and litigation from the Foundation on Economic Trends, Jeremy Rifkin's organization based in Washington, D.C.

Between 1980 and 1985 the RAC reviewed five proposals for field-testing GEMs. Four were approved and one was voluntarily withdrawn (see Table 6.1).

GENETICS AND THE ENVIRONMENTALISTS

In the mid-1970s, when rDNA molecule research was still an exotic technique, the public entry into the debate over its safe uses was determined by several

Table 6.1
RAC Proposals Involving Field-Testing of GEMs, 1980–1985

Proposal	Submitter	Date of Submission	Date of RAC Review	Date of Approval
Corn plants with corn DNA added that was cloned in E. coli or S. cerevisiae.	Stanford U.	2/9/80	6/6/80	8/7/90
Tomato and tobacco plants with bacterial and yeast DNA.	Cornell U.	6/9/80	12/25/82	4/15/83
Pseudomonas syringae and Erwinia herbicola with ice-nucleation genes deleted.	U. Cal. at Berkeley	9/17/82 4/11/82	10/25/82	6/1/83
Tobacco plants grown from seeds with rDNA for resistance to crowngall disease.	Cetus Madison	5/5/83	9/19/83	11/13/85
Pseudomonas strains with ice-nucleation genes deleted.	Advanced Genetic Sciences	3/22/84	6/1/84	Voluntarily Withdrawn

Source: Elizabeth Milewski, "The NIH Guidelines and Field Testing of Genetically Engineered Plants and Microorganisms," *Application of Biotechnology: Environmental and Policy Issues*, John R. Fowle, III, ed. (Boulder, CO: Westview Press, 1987), 55–90.

factors. First and foremost, a group of distinguished biologists called attention to the new technique by publishing letters in science journals that caught the attention of the popular press. The international meeting at Asilomar was attended by science writers, whose coverage of the events dramatized the disagreements among biologists at the meeting. In drawing attention to the new discovery, the Asilomar organizers did not intend for the fate of the new technology to be decided by popular acclamation or democratic process. However, some biologists, frustrated by the NIH's role in setting safety standards, brought their concerns to university campuses and local communities. To the public, it appeared that scientists were polarized on the dangers of rDNA research.

Second, the issues were dramatized through the world press by the controversy that erupted in Cambridge, Massachusetts, which pitted community values against the interests of national science.[19] The 1976 Cambridge rDNA controversy represented the birth of public involvement in genetic engineering. Prior to that, the issues were debated exclusively in professional groups and on university campuses, notably the University of Michigan.

Third, national environmental groups such as Friends of the Earth, the Environmental Defense Fund, and the Sierra Club devoted some attention to genetics issues between 1976 and 1979. But the commitment by these groups was limited usually to a single staff person. Their activities were met with skepticism, and in some cases, disapproval by some board members who had ties to the biomedical community. Fourth, but to a lesser extent, litigation had drawn public attention to the issues. Two suits were filed against HEW/NIH in 1977. A resident of Frederick, Maryland, petitioned the court to enjoin the NIH from performing a risk assessment experiment until an environmental impact statement was prepared. A second suit filed by Friends of the Earth alleged that HEW failed to comply with the requirements of the National Environmental Policy Act when it established policies for rDNA research.

Fifth, congressional interest in the safety of rDNA research intensified in the mid-to-late 1970s. House and Senate hearings provided a forum for activist scientists and environmental lobbyists. The accumulated effect of these factors resulted in heavy media coverage for both local and national events related to gene-splicing experiments.

The subsequent public debate over the specific issue of the deliberate release of genetically engineered organisms has been fostered by a similar configuration of factors, namely, scientist-critics, litigation, environmental coalitions, local opposition to experiments, and congressional interest. However, the old and new rDNA debates differ in the weighting of these factors. For example, Jeremy Rifkin has proven particularly effective in drawing public attention to the environmental and ethical issues associated with the deliberate release of GEOs. Operating with the modest resources of his nonprofit organization, Rifkin has utilized a variety of techniques, including confrontational protests, litigation, and short-term coalition building, to dramatize breaking issues in genetics and public policy.

Rifkin's philosophical opposition to genetic engineering has been expressed in his writings and media accounts that feature his work: "Genetic engineering represents the ultimate negation of nature."[20] On genetics policy, he has had more impact on the media that any single group or individual in the United States. His notable success as a publicist and genetics critic can be explained by several factors. Rifkin can act quickly since he is not accountable to a board of overseers or a mass organization. He is able to identify and capitalize on the weakest link of any policy process. With properly timed and targeted litigation, Rifkin has been able to strike at the jugular of the bureaucracy. He has proven time and again that a well-placed lawsuit is a magnet for media attention. Rifkin made his national debut as a genetics activist in 1977 when he and a group of supporters invaded a meeting at the National Academy of Sciences devoted to rDNA science and policy. His strategy at the NAS is best described by the 1960s term "guerrilla theater," a dramatically staged political confrontation. Fearing major disruption of its conference, academy officials allotted Rifkin time at the start of the program for a statement.[21] Meanwhile, his associates, donning silk stocking face masks, stood like motionless icons, projecting a grotesque apparition of human mutants. While Rifkin took over the podium of the academy auditorium, his colleagues stretched a banner across the stage with a quote from Adolf Hitler that stated: "We will create the perfect race."

On another occasion, Rifkin organized a signature campaign of scientists and

religious leaders who were asked to support a statement against genetic manipulation of the human germline (see Chapter 9). The list of supporters included
individuals of vastly different ideological persuasion. To have such a politically
diverse group on the same side of an issue was itself a news event. It proved
worthy of a front-page story in the *New York Times*.

In the fall of 1983, when the RAC was reviewing a proposal for field-testing
ice minus, Rifkin brought national attention to the problem of deliberate release
by filing suit against NIH on the grounds that its approval was given prior to a
full study of the environmental consequences. His suit was supported by two
organizations, Environmental Action and the Environmental Task Force. To
strengthen his case, Rifkin received sworn affidavits from leading ecologists who
opposed the release of the ice minus strains, at least until more could be learned
about the environmental consequences.

Rifkin's lawsuits have been rallying points for critical debate on genetics
issues. Philosophically, he personifies the modern genetics Luddite. But on a
practical level, he has made effective use of environmental law to impose a
greater burden of responsibility for assessing the risks and eliciting the ethical
consequences of new technologies. Ironically, he is secretly applauded by many
who, while they might differ with him philosophically, find a certain degree of
relief in slowing the pace of regulatory approval. Environmental groups, skeptical
about deliberate release, have been hesitant to act more aggressively because
the hazards of genetic engineering are still hypothetical. For Rifkin, it was less
important to prove the hazards exist than to rally support against the hubris of
science and reexamine its assumptions about progress.

There is a striking comparison between the old and new genetics controversies
on the role of scientific participation. In 1976, the molecular geneticists were
taken to task for not calling upon infectious disease epidemiologists to evaluate
the risks of transforming *E. coli* into an epidemic pathogen. Subsequently, the
infectious disease community did participate in the risk evaluation. Similarly,
ecologists played no role in the rDNA debate until the controversy over deliberate
release erupted. Their entry into the discussion over deliberate release came
when the social and political context was ripe for their participation. Ecologists
are viewed as natural allies to environmentalists; when the latter became involved
in the debate over field tests, they sought advice from the former. Also, the
Committee on Science and Technology of the House of Representatives made
specific recommendations to NIH and USDA that they revise the memberships
of their advisory committees to include individuals specifically trained in ecology
and environmental sciences.[22]

Finally, the EPA began developing its own policy on deliberate release in the
early 1980s. The agency is accustomed to working with ecologists, for example,
in evaluating pest control methods. The first genetically engineered organisms
planned for environmental release were classified as biological control agents.
The EPA began hiring consultants to study the potential environmental problems
of releasing such agents. A terrestrial ecologist from Oak Ridge National Lab-

oratory, Frances Sharples, completed a study for the EPA in 1982 on the effects of the introduction of organisms with novel genotypes into the environment. Microbial ecologist Martin Alexander of Cornell University chaired an EPA study group on biotechnology. In 1984, ecologists participated in an EPA-sponsored workshop at the Banbury Center at Cold Spring Harbor, New York, on the evolutionary consequences of biotechnology. Also, in 1985, with support from the EPA, the USDA, national environmental organizations, and corporations, ecologists, geneticists, public interest representatives, and federal regulators met for four days at Cornell's biological field station in Bridgeport, New York, to develop principles of containment for genetically engineered organisms tested under field conditions.[23]

By raising the issues of biotechnology in the national media, Rifkin also helped catalyze the concerns and perspectives of ecologists on the problems of deliberate release, particularly those who had not been brought in as consultants to the major agencies. Eventually, letters from ecologists appeared in *Science*,[24] followed by responses from geneticists.[25] A dialogue was opened between molecular geneticists, microbiologists, and ecologists on their approaches to risk assessment.[26] Soon it became evident that the issues over ice minus unfolded into a larger debate about scientific culture, epistemology, and disciplinary hegemony.

The next chapter looks at how the first field test of a genetically engineered microorganism was evaluated by the EPA and received by the general public and citizens in communities slated for the tests. We will see how the ice minus bacterium became an international symbol for resistance against biotechnology by some groups like Earth First and the European Greens. To the biotechnology industry, ice minus had quite a different symbolic value as a product that could raise public confidence in a new technological frontier that promised to advance agricultural efficiency while making peace with nature.

7

Controlling Frost with Bacteria:
The First Field Test

The Deliberate Release of "Ice-Minus" would be a grave, yet maybe ir-reversible hazard to the environment. The risk and threat of unforseen con-sequences would affect the beautiful landscape of Monterey County.
                                        Members of the European Parliament in a telegram
                                        to the Monterey County Board of Supervisors, 1986

AGS [Advanced Genetic Sciences] has conducted more than 200 tests on Frostban. . . . Never before has a technological advancement for agriculture been so thoroughly tested and reviewed prior to its first field trials.
                                        Advanced Genetic Sciences, from a public relations videotape, 1986

In *Cat's Cradle*, Kurt Vonnegut created a satiric fantasy that poked fun at scientific elites, ideologues, and American culture. The story describes a sub-stance called ice nine that has the combined properties of an infectious agent and liquid nitrogen. Ice nine is deadly to biological life. It is capable of slowing down atomic processes and transforming living cells into functionless ice-like forms. The substance eventually gets out of control and threatens civilization.

Von Koenigswald touched the tip of his tongue to the blue-white mystery. Frost bloomed on his lips. He froze solid, tottered, and crashed. The blue-white hemisphere shattered. Chunks skittered over the floor. . . . At last I had seen ice nine![1]

There is nothing more than a symbolic connection between Vonnegut's ice nine that transforms living matter to a crystalline form and a genetically engi-neered bacterium called ice minus that was developed to inhibit ice crystalli-zation. But it is one of those coincidences that draws attention to the shifting boundary between science and fantasy. The mystery of science is capable of stimulating the literary imagination while also nourishing public anxiety about technology.

Since the Enlightenment, science has contributed to a rich tapestry of cultural myths, metaphors, and worldviews. It continues to have a profound impact on literature. But fiction has also influenced social attitudes toward science. Mary Shelley's gothic novel *Frankenstein* gave us a myth that still holds an important place in folk consciousness. During the 1976 genetic debates in Cambridge, Massachusetts, the "Frankenstein factor" loomed prominently in the public discourse. Another popular novel by Michael Crichton published in 1969 and titled the *Andromeda Strain*, provided another metaphor through which laypeople and some scientists viewed the "awesome" possibilities of genetic engineering.[2] The proliferation of grade B movies about mutant organisms invading the environment has also implanted powerful imagery in the minds of the popular culture.

The public understanding of science is filtered through the fanciful imagination of the cinema and works of science fiction. But there are also times when science appears more like fiction than truth. Some results seem to defy our intuitive grasp of the order of nature. That is one of the important lessons of the ice minus episode.

Most people understand that when the temperature drops below freezing, their tomato plants will die unless some action is taken to protect them. But who can imagine spraying a microbe on the plants to prevent the onset of ice crystals? What does the formation of ice have to do with bacteria? What might have been a plausible story line for a science fiction novel proved to be a serious research program in plant genetics.

Scientists developed a microbe that they believed could reduce the temperature at which ice crystals would form on vegetation. It was the first genetically modified bacterium sanctioned by a federal agency for release into the environment and it sparked a national debate. The story of ice minus illustrates the interplay of science and human imagination. First, it was the scientific imagination that established the causal connections between bacteria and frost. And second, it was the public imagination that brought world attention to the consequences of tampering with the earth's microflora.

The discovery of ice minus dramatizes to us that the natural world operates in unexpected and often counterintuitive ways. Nature scoffs at a Cartesian mechanism. Physical, chemical, and biological states relate through complex webs of interdependent and interpenetrating processes. Boundaries between disciplines are what we impose. They are not nature's design. Perhaps because of its mystery and open-endedness, science has become a vast reservoir for the literary imagination. If ice minus were not a product of the laboratory, it might well have appeared in the new genre of biofiction. With Vonnegut's "ice nine" it came very close.

DISCOVERY OF ICE MINUS BACTERIA

The story of ice minus begins around 1974 when scientists discovered that certain soil bacteria, such as *Pseudomonas syringae* and *Pseudomonas fluores-*

cens, efficiently catalyzed ice formation at temperatures several degrees below zero centigrade.[3] Water does not always freeze when it is cooled below zero. It can remain in a liquid (supercooled) state when ice-nucleating particles are not present. However, when ice nuclei are present in the form of inert particles of matter or bacteria, then freezing takes place at or very near zero degrees.

Plants are damaged by extracellular ice that forms within their tissues. Ice crystals mechanically disrupt biochemical processes. If ice nuclei are absent when the plant is exposed to moisture and subfreezing temperatures, water will supercool. The supercooled water is capable of remaining in a liquid state at temperatures that approach $-40°C$. Thus, frost injury to plants requires low temperatures, moisture, and ice nuclei.

It has been estimated that about 11 percent of the land mass of the continental United States is allocated to the cultivation of frost-sensitive plants. Losses to crops from frost injury is reputed to be about $1 billion annually.[4]

The agricultural sector has experimented with various techniques to minimize the impact of frost on important crops, particularly in those regions where freezing temperatures, within the subzero range of ten degrees, occur for short periods during critical stages of the growing season.

Traditional methods of protecting crops from frost damage are premised on keeping the air temperature above freezing in the vicinity of the harvest. Farmers have resorted to burning smudge pots, watering the soil, utilizing wind machines to keep warmer air circulating, and applying foam-like insulation around the plants. Reductions in frost injury under field conditions have also been achieved through the application of bactericides that are used to destroy ice-nucleation microbes.[5]

The discovery of gene-splicing provided a novel approach to protecting plants from frost damage. It involved replacing the indigenous ice-nucleation bacteria with strains that have had the ice-nucleation genes excised. Pseudomonas syringae is a ubiquitous organism that resides on the leaf surfaces of plants. The role of this bacterium in the agricultural ecosystem has been studied since 1902 when the organism was first isolated.[6] In the early 1970s, scientists learned that this organism was unusually efficient at seeding ice crystals from subfreezing water. Most organic and inorganic substances that are ice-nucleating agents (INA) are active below temperatures of $-10°C$. P. syringae, however, catalyzes the formation of ice crystals at temperatures between $-1.5°C$ to $-5°C$. This characteristic of the microbe attracted special interest from plant pathologists who study the physical environmental and biological factors that impede plant growth.[7]

The mechanism of ice nucleation for P. syringae has been traced to the bacterial synthesis of a protein. The release of this protein provides a site for crystallization of moisture as the temperature drops below zero. Strains of P. syringae that lack the protein fail to exhibit ice-nucleating activity at the higher subzero temperatures.

About one half the strains of P. syringae that have been tested are ice-nucleating agents. Also, through chemical mutagenesis, scientists have been able to transform an INA *P. syringae* into a non-INA type. One method devised to reduce crop damage during subfreezing temperatures is to colonize the seedlings of a plant with non-INA bacteria. If the antagonistic bacteria take over the niche of the INA type, frost formation could be delayed, at least until the temperature drops below $-10°C$. This process requires displacing about 0.1 to 10 percent of the bacteria on leaf surfaces since that represents the percentage of ice-nucleating strains.[8]

Steven Lindow of the University of California (UC) at Berkeley carried out some of the pathbreaking work in studying the relationships between INA bacteria and frost formation on plants. He discovered how bacteria can limit the damaging effects of moisture to frost-sensitive plants at temperatures above $-5°C$. He also demonstrated that by reducing the numbers or ice-nucleation activity of the INA bacteria, frost injury to field-grown plants declined.[9]

Although the number of INA species is relatively small, the prevalence of the bacteria in the environment is quite high, particularly on certain food crops. *P. syringae* are widely distributed in nature and are believed to be one of the largest sources of bacterial ice nuclei that are active at high subzero temperatures (near zero degrees). By reducing the population density of INA bacteria, scientists believed they had an effective and environmentally safe method of protecting crops from sudden frosts. With an estimated billion dollar annual loss to agriculture from frost damage, there was a strong financial incentive to develop the new technology.

COMMERCIALIZING A DELETION MUTANT

The Oakland, California, biotechnology firm Advanced Genetic Sciences (AGS) initiated an R&D program to commercialize a microbial frost inhibitor. Of the several possible approaches for developing antagonistic bacteria to displace INA *P. syringae*, the chosen path was to modify the microbe by removing the ice-nucleating gene, producing a so-called deletion mutant (denoted as INA⁻). Initially, AGS accomplished this by exposing INA⁺ (the strain with the ice-nucleating gene) to chemical mutagens. The chemical mutant was produced in the laboratory and field-tested. Special permits were not required for these field tests since the product was not a genetically engineered microbe. AGS reported that the INA⁻ produced by mutagenesis slowed down the effects of frost damage. But the company believed it could improve the effectiveness of the product.

AGS then deleted the ice-nucleation gene from *P. syringae* using recombinant DNA techniques.[10] In the spring of 1984, the company submitted a proposal to the RAC for field-testing the genetically engineered *Pseudomonas* strain. The proposal enunciated that the rDNA deletion mutants of *P. syringae* (INA⁺) are more stable, better defined genetically, and probably more fit than the INA⁻ strain resulting from chemical mutagenesis. In its submission to the NIH committee, the company disclosed that it had removed about 1,000 base pairs of DNA sequences and that the changes were nonrevertible.[11]

AGS maintained that the rDNA mutant strains were, from the viewpoint of ecological risk, no less safe than the strains isolated in nature or produced by chemical mutagenesis. Furthermore, the company contended that the rDNA-derived strain was superior to its chemically derived counterpart in inhibiting frost damage. Thus, the natural and genetically modified strains were considered alike in all respects with regard to risk, but different in some respects pertaining to the organism's ecological role. Meanwhile, industry spokesmen and some government officials pointed to the hypocrisy of regulating a process and not a product. If the product is potentially hazardous, they argued, it should not matter whether it was derived by chemical mutagenesis, gene-splicing, or cultured from nature.[12]

NIH REVIEW OF ICE MINUS BACTERIA

The National Institutes of Health, through its Recombinant DNA Advisory Committee, had been overseeing gene-splicing experiments since 1976, when the first guidelines were issued. Its initial involvement in reviewing ice minus began in September 1982 when Steven Lindow and Nicholas Panopoulos (both at UC Berkeley) requested approval to field-test two strains of INA-modified bacteria, *P. syringae* and *Erwinia herbicola*, that had all or part of the genes for ice-nucleation deleted. At the time, no other government body had exercised any regulatory authority over the deliberate release experiments.

The RAC reviewed the proposal on October 25, 1982. Several concerns about ice minus were expressed by committee members. Among them were: (1) Was there sufficient data on these organisms? (2) Why should the field test take place at six different sites? (3) Could release of the organisms affect rainfall patterns in California? Despite some unresolved issues expressed by several committee members, the RAC approved the field test, albeit by a divided vote (seven in favor, five opposed, and two abstentions).

Decisions of the RAC constitute recommendations that are then reviewed by the director of the NIH. Split votes have generally drawn conservative responses from the director who, ordinarily, was briefed by a small group of rDNA advisers informally known as the "kitchen RAC."[13] The recommendation to field-test the INA strains was rejected by the director of the NIH (January 10, 1983) pending submission of additional laboratory data.

The UC Berkeley proposal was revised, resubmitted, and reviewed by the RAC in April 1983. At that time the committee was satisfied that the Berkeley scientists had met previous objections. To monitor the organisms in the field, the strains would be marked with resistance factors for non-clinically used antibiotics. In this second review, the RAC unanimously approved the proposal to field-test the ice minus strains. The proposal was then submitted to the Department of Agriculture's (USDA) recombinant DNA adivsory committee, which also gave it the green light. The final approval by the NIH was formally announced in the *Federal Register* in June 1983.[14]

Trailing close behind UC Berkeley, Advanced Genetic Sciences submitted a proposal to the RAC for field-testing an INA⁻ strain in March 1984. The firm outlined the safety precautions it was taking for the planned field experiment and cited as a primary safety consideration that any dispersal of bacteria beyond the test plot would not be in sufficient amounts to colonize plants. AGS pointed out that the INA⁻ does not survive well through a full year cycle, suggesting that its use in a field plot will be limited both in space and time. AGS also included in its submission packet to the RAC an essay by Steven Lindow that discussed the possible effect INA⁺ bacteria might have on atmospheric precipitation processes: "Yet the importance of ice nucleation-active bacteria in contributing ice-nuclei active at warm temperatures in the upper atmosphere for the formation of rain and snow is as yet largely unexplored. The role of these bacteria may be potentially of critical importance in climatology studies."[15]

The firm could hardly neglect this risk consideration cited by one of the world's experts on INA bacteria. AGS acknowledged that "naturally-occurring INA⁺ bacteria may be an important factor influencing precipitation processes." But, in its written proposal, the firm maintained that field trials could not significantly reduce the level of bacteria in the atmosphere. No analytical calculations or atmospheric modeling were part of the submission materials.

In another argument, AGS asserted that the INA bacteria proposed for release are genetically equivalent to simple deletion mutants that arise spontaneously in all bacteria. They sketched out a model of bacterial ecology according to which INA⁻ are continuously evolving from INA⁺ strains through a natural mutation process. Since the spontaneous INA⁻ are genetically identical to the laboratory-produced strains, they argued that "in this field test the environment is not being used as a testing ground for a 'new biological product' or a 'novel genotype.' Natural strains of ice minus and ice plus cohabit in nature, therefore adding a little more of what is already there in plentiful supply could have no adverse consequences."[16]

Pathogenicity was another factor considered in assessing the risk of ice minus. According to Lindow, most isolates of *P. syringae* are pathogenic on some plant species. However, for the field test, he chose a strain that did not exhibit pathogenicity to any of a large list of plants that were potential hosts, including all major agricultural crops in northern California.[17]

The assessment procedure established by NIH for deliberate release experiments involving rDNA-modified organisms and plants incorporated a multitiered review that included an institutional biosafety committee, the RAC, panels in relevant agencies, and the director of NIH. For certain genetically engineered plants, an accelerated review process was initiated that required approval by NIH's Office of Recombinant DNA Activities in consultation with the RAC plant working group. In the accelerated review process there was no requirement for approval by the full RAC.[18] Microorganisms, however, did not fall into this category.

During this period, the NIH's risk assessment placed primary emphasis on molecular genetics, relegating ecology to a mere shadow of significance. The two primary considerations in the RAC review were whether the foreign genes might express a dangerous gene product and whether those genes might be transferred to a species related to the host organism.

From its inception, the RAC's composition was dominated by molecular geneticists who have a professional interest in *in vitro* gene transfer and the biochemical anatomy of organisms. Far less significance was given to the effects a new species might have on different components of the environment. Included among these effects are alterations in nutrient cycles and broad interactions among microorganisms, plants, and animals. To illustrate the distinction between a genetic and an ecological approach to the problem of deliberate release, consider the following reconstructed argument of a hypothetical molecular geneticist. A well-characterized nucleotide sequence is introduced into a plant that is not considered a weed. Let us assume the foreign sequence is not known to be associated with any noxious weeds. From the genetics perspective, it can be assumed that a weed would not result from transplanting the foreign genes. Similarly, if the foreign sequence is transferred naturally to a weed unrelated to the host plant, one can say with reliability that the new genetic sequence will not enhance the weed-like characteristics of the recipient species.

In simple terms, the genetics approach to risk assessment is reductionist. It assumes that an a priori evaluation of the genetically modified product can be made by an empirical assessment of the starting materials. The guiding principle is: from non-weeds, weeds do not result. No room is left for emergence of new properties.

Ecologists were critical of such assumptions. They expressed skepticism about predicting the properties of the whole from the properties of the parts. Ecological theory had not advanced far enough to make such forecasts. Moreover, some ecologists did not believe that macrosystem effects could ever be explained from the genetic level. A more detailed discussion of the distinctive approaches of geneticists and ecologists in evaluating the risks of GEOs is given in Chapter 8.

RIFKIN'S INTERVENTION

Criticism of the RAC's assessment of the ice minus field test came from several sources, including independent environmental groups, ecologists, and Jeremy Rifkin. A report from the centrist Environmental Law Institute (ELI) identified fundamental weaknesses in the RAC's approach. Writing for the ELI's *Environmental Law Reporter*, Frances McChesney and Reid Adler questioned NIH's approval of the deliberate releases without issuing generic standards.[19]

The authors characterized the NIH's risk assessment as giving greatest weight to the expression of dangerous gene products and the transfer to other species of the foreign genes. These issues are the most easily reconcilable within a framework of molecular genetics. McChesney and Adler faulted the RAC for not posing the difficult questions.

Widespread dissemination of "anti-freeze" bacteria could produce harmful outcomes as well, but the likelihood is impossible to estimate. The scientist to whom permission for the experiment was granted admitted that "it's hard to tell for sure" what effect on global weather might be caused by the presence of the modified organisms once carried into the upper atmosphere.[20]

The critic's attention was drawn to two issues. First, there is the impact that the widespread use of ice minus would have on precipitation; second, there are the unarticulated and unanticipated consequences the organism might have on the ecology of the region.

With support from two environmental organizations, Jeremy Rifkin initiated a suit against the NIH in September 1983. The legal brief requested that the court enjoin the NIH from approving deliberate release experiments of genetically modified agents. Rifkin's philosophical and historical analysis of modern biology has gained little respect or recognition among scholars. Stephen Jay Gould's review of *Algeny* stands out as the most significant commentary from the intellectual community on Rifkin's lack of scholarly integrity: "I regard *Algeny* as a cleverly constructed tract of anti-intellectual propaganda masquerading as scholarship. Among books promoted as serious intellectual statements by important thinkers, I don't think I have read a shoddier work."[21] In stark contrast, his carefully researched and well argued legal briefs have tested the boundaries of environmental law and brought the issues around biotechnology to center stage in the public policy arena.

In his suit against the NIH, Rifkin argued that the agency failed to apply a set of reasonable protocols for estimating the survivability and growth of the experimental organisms prior to field testing. He also noted, correctly, that the RAC included no ecologists, botanists, plant pathologists, or population geneticists, and thus lacked important areas of expertise in rendering a judgment about risks. In addition, Rifkin cited as a procedural violation under the National Environmental Policy Act—the NIH's failure to issue an environmental impact statement for the field tests. The NIH responded to the criticism that the RAC's composition was not representative of disciplines relevant to a review of deliberate release by adding ecologist Frances Sharples to the RAC.

Rifkin eventually won a federal court injunction against the NIH. Judge John Sirica handed down a decision in May 1984 that placed a moratorium on all field tests of genetically altered organisms approved by the NIH. Moreover, the court order stopped the NIH from approving any more experiments until the suit was settled. As a result, the Lindow-Panopoulos field test scheduled for May 25 was postponed.

In February 1985, the U.S. Court of Appeals of the District of Columbia lifted the preliminary injunction provided the NIH issue an environmental impact assessment for deliberate release. The procedural actions taken by Rifkin and supported by the courts did not reveal new insights about the ecological effects of genetically engineered microbes. The agencies generally repackaged the information they had already amassed. But those legal skirmishes slowed up the process of approving new tests and ultimately of getting products to the market. Regulators expressed frustration over the legal challenges; this was particularly the case at the EPA, where efforts were made to establish an exemplary review process for ice minus.

Rifkin succeeded in becoming the czar of genetics litigation. As issues arose in biotechnology that troubled environmentalists, the prevailing attitude among leading activists who were knee deep in toxics issues was that the Foundation on Economic Trends (FET) will probably litigate. And more often than not they were correct. Despite a modest staff, the foundation had an impressive record of legal challenges that included cases on transgenic animals, deliberate release, human genetic engineering, and the military's biological defense weapon's program (see Table 7.1).

Perhaps Rifkin's most successful legal challenge in a half-dozen years was filed in 1984 to stop construction of a biological research aerosol laboratory at the Dugway Proving Grounds in Dugway, Utah. The plaintiff argued that an accidental release of deadly viruses could devastate local livestock and inflict harm to human populations. In May 1985, a federal district court judge ruled that the risks of the $1.4 million "aerosol test lab" were "serious and far-reaching," that the Army's 1984 environmental assessment was inadequate, and therefore the program under consideration represented a substantial violation of the National Environmental Policy Act.[22] FET was granted a permanent injunction against construction of the laboratory. More significantly, the litigation sparked local protests and coalitions between environmentalists, public health advocates, and disarmament groups. Under these conditions, there is doubt that the facility will ever be built.[23] In September 1988, DOD scaled down its plan for building a high-containment facility (level 4) to a lower-containment facility (level 3), where presumably there are restrictions against testing the deadliest human pathogens: "Army spokesmen acknowledged that the mounting opposition to its plans from local citizens, scientists, and politicians was 'a factor' in the decision to downgrade the facility."[24]

Among leading environmental groups in the United States, there was moderate interest but little capacity to undertake legal actions similar to those of Rifkin's organization. It was a question of priorities. Superfund, acid rain, contaminated water supplies, depletion of genetic resources, global warming, pesticide residues in food, and conservation of fragile ecosystems were high on the agenda of environmental groups. Biotechnology was still new and its risks were hypothetical. Nevertheless, environmentalists recognized that there was an opportunity to regulate a technology before a hazard appeared. But in the bizarre world

Table 7.1
Rifkin's Lawsuits in Biotechnology, 1983–1987

1. Filed: Sept. 14, 1983.   FET v. HECKLER   No. 83-2714 D.D.C.
   Release of Ice Minus Bacteria.
   The FET petitioned the court to stop plaintiffs (DHHS, NIH and the University of California) from releasing genetically-engineered ice minus bacteria. Decided: May 16, 1984.

2. Filed: Oct. 1, 1984.   FET v. LYNG   No. 84-3045 D.D.C.   86-5452 (appeal).
   Genetically-engineered Animals.
   The FET challenged the Department of Agriculture for its role in approval of genetically-engineered farm animals designed for increased productivity and accelerated maturation. Decided: April 29, 1986; May 1, 1987 (appeal).

3. Filed: Nov. 21, 1984.   FET v. WEINBERGER   No. 84-3542 D.D.C.
   Dugway Biological Defense Laboratory.
   The FET petitioned the court for a permanent injunction against the Department of Defense's construction of a biological research aerosol laboratory (BL-4) and challenged the U.S. Army's failure to conduct a satisfactory environmental assessment of the proposed facility. Decided: May 31, 1985.

4. Filed: Nov. 4, 1985.   FET v. BLOCK   No. 85-3510 D.D.C.
   Germ Plasm Resources.
   The FET challenged the USDA for not carrying out an environmental assessment of its germplasm conservation program. Decided: Nov. 22, 1985. The USDA agreed to conduct an environmental assessment of its program.

5. Filed: Nov. 14, 1985.   FET v. THOMAS   No. 85-3649 D.D.C.
   Permit to Field Test Ice Minus.
   The FET challenged the EPA's proposed approval of an experimental use permit (EUP) under FIFRA for a genetically-engineered ice minus bacterium developed by Advanced Genetic Sciences of Oakland, California. Decided: March 6, 1986.

6. Filed: April 23, 1986.   FET v. LYNG   No. 86-1130 D.D.C.
   Pseudorabies Vaccine Release.
   The FET challenged the Department of Agriculture's approval of an animal bovine pseudorabies vaccine without having conducted a suitable environmental assessment prior to the release of the vaccine into the environment. Decided: Jan. 11, 1988.

7. Filed: June 9, 1986.   FET v. THOMAS   No. 86-1590 D.D.C.
   Liability Requirements for Deliberate Release Tests.
   The FET challenged the EPA for failing to develop regulations requiring that companies seeking to release GEOs into the environment under the FIFRA demonstrate financial responsibility for any adverse consequence. Decided: Dec. 22, 1986.

Table 7.1 (continued)

8. Filed: July 15, 1986.   FET v. JOHNSON   No. 86-1956 D.D.C.
   Coordinated Framework for Regulating Biotechnology.
   The FET challenged the White House and six federal agencies for the procedures it used in issuing the Coordinated Framework for the Regulation of Biotechnology and for the substantive content of the framework itself. Decided: Dec. 22, 1986.

9. Filed: Aug. 1, 1986.   FET & CALIFORNIANS FOR RESPONSIBLE TOXICS MANAGEMENT v. REGENTS OF UNIVERSITY OF CALIFORNIA No. 342097 (CA)
   Ice Minus Release.
   Plaintiffs sought a restraining order prohibiting the University of California from releasing the first genetically-engineered microbe in Tulelake, California. Decided: April 1987.

10. Filed: Sept. 2, 1986.   FET v. WEINBERGER   No. 86-2436 D.D.C.
    Biological Defense Development Program.
    The FET challenged the DOD for violations of the National Environmental Policy Act in the conduct of its Biological Defense Research and Development Program. Decided: Feb. 12, 1987.

11. Filed: Oct. 27, 1987.   FET v. LYNG   No. 87-2909 D.D.C.
    USDA's Environmental Assessment.
    The FET challenged the USDA on the adequacy of its environmental assessment for its germplasm conservation program. Decided: ?

12. Filed: Dec. 15, 1987.   FET v. BOWEN   No. 87-3393 D.D.C.
    AIDS Mouse.
    The FET challenged the NIH for the creation of an AIDS mouse by implanting the AIDS virus gene into a mouse embryo. Decided: ?

of environmental politics, one cannot underestimate the importance of dramatizing a risk as a means of building coalitions and gaining broad public support. The potential for dramatizing risks in biotechnology, where there have been no "Bhopals" or "Chernobyls," was limited unless one was willing to exercise a bit of imagination and hyperbole. An alternative way to engage the media and ultimately the broader public was to draw attention to procedural violations and weaknesses in the regulatory process. Rifkin has been masterful at both these techniques.

In addition to his litigation, Rifkin also brought public attention to irregular or misguided practices of government and business. Whistle blowers, familiar with Rifkin's track record as a publicist, viewed him as a reliable channel to the media. A *New York Times* editorial cited Rifkin's role in publicizing the "deceit and recklessness of some early practitioners" in biotechnology. There were three cases of note between 1986 and 1987. The first case involved unauthorized tests by Advanced Genetic Sciences. An employee of AGS contacted Rifkin through a lawyer and disclosed information about releases of a strain of ice minus bacteria in trees situated on the roof of the firm's facility.

The second case concerned the USDA's Animal and Plant Health Inspection Service's approval for testing a genetically altered herpes virus for pigs. The release of the herpes virus was not reviewed according to internal procedures established within the USDA for rDNA-derived organisms. The *Times* gave Rifkin a resounding vote of confidence for public interest work.

The Foundation on Economic Trends, an unremitting watchdog of genetic engineering, deserves credit for bringing to light these two cases and what they show about the ragged Federal system for regulating the new technology. The system fails to protect the public—and its delays and inconsistencies sorely try the new industry.[25]

In the third case, the Wistar Institute carried out field tests of a recombinant rabies vaccine in Argentina during the summer of 1986. Neither Wistar or its sponsor, the Pan American Health Organization, notified the Argentine or U.S. governments that the test was taking place. Argentine scientists criticized the careless execution of the experiment, citing poor medical surveillance and a breakdown in the protocols.[26]

The publicity around events like these stimulated congressional interest in a stronger regulatory posture over biotechnology. The NIH's role in setting up a voluntary compliance program for industry was a default measure that could not oversee the complex operations of private-sector activity. The next logical step in a biotechnology policy that some would characterize as a policy of "disjointed incrementalism," was to broaden the EPA's role and retrofit biotechnology to the existing statutes.

THE EPA TAKES ON BIOTECHNOLOGY

When Reagan took office in 1981, his administration began an aggressive policy of deregulation. His neo-conservative philosophy of reducing government's role in the social and commercial affairs of life was sometimes derailed, however, by public advocacy, resistance within the bureaucracy, or congressional pressure. As part of this counterwave, in the early 1980s Congress held hearings on the deliberate release of genetically engineered organisms in the wake of lawsuits and media allegations of regulatory deficiencies. The EPA was urged to create a public policy for both field tests and large-scale use of genetically altered life-forms. During that period, the EPA administrator Ann Burford Gorsuch was redirecting the agency away from vigorous implementation of the Superfund law and other regulations involving hazardous waste and toxic substances. Gorsuch's head of toxic substances, Rita Lavelle, former director of communications for the Aerojet-General Corporation, developed a sweetheart relationship with industry that created intense dissension within the agency and eventually led to her conviction and Gorsuch's resignation.

Under the new leadership of William Ruckelshaus, the EPA took some steps toward developing a biotechnology policy in the midst of funding cuts, demoralization from within, and a severely tarnished public image. In August 1983, while the NIH was reviewing industry proposals for deliberate-release experiments, the EPA asserted its regulatory authority over products in biotechnology under the Federal Insecticide, Fungicide and Rodenticide Act (FIFRA) and the Toxic Substances Control Act (TSCA).

One area of biotechnology where the EPA had made some headway was in the regulation of microbial pesticides. The evaluation and licensing of microbial pesticides in the United States predates the environmental movement and the EPA by several decades. The first microbial pesticide, *Bacillus thuringiensis*, was registered in 1948. By 1985 there were 14 non-genetically engineered microbial pest-control agents registered in over 100 products. The agents are used in agriculture, forestry, and for mosquito control. In contrast to chemical alternatives, microbial pesticides are beneficial because of their specificity to insects and their low toxicity to humans and other mammals.[27]

The EPA's guidelines for microbial pesticides took a number of years to develop, but were not written with genetic engineering in mind. The agency took additional precautions in reviewing viral and bacterial agents that had undergone genetic modification. In December 1984, the EPA developed an interim policy on small-scale field-testing for microbial pesticides. Small-scale was defined as the application of a substance on ten acres of land or less.

Under the policy, the EPA must be notified prior to such a test and has 90 days to review a proposal. If the agency decides that the proposed test raises health or environmental concerns, an Experimental Use Permit (EUP) may be required before testing can begin. The policy for field-testing GEMs did not apply to studies for which there was no anticipated direct release to the environment, such as with growth chambers, contained facilities, and greenhouses.[28]

By 1985 the EPA had received five notifications for field tests. Two proposals involved INA⁻ bacteria; another involved the insertion of a toxin gene from *Bacillus thuringiensis* into a soil bacterium; the fourth involved test strains of a fungus altered by undirected mutagenesis using ultraviolet irradiation and chemical treatment. A fifth proposal was withdrawn. One of these submissions came from Advanced Genetic Sciences, which sought EPA approval to field test its INA⁻ deletion mutant.

The EPA had put into place what it considered to be an exemplary review process that involved several stages of consultation, including interagency working groups and a special ad hoc subpanel of its external advisory group. With all the information in hand, the EPA concluded that the AGS tests appeared to pose a low likelihood of adverse environmental or human health effects. However, sensitive to the public attention directed at the issue, the agency requested additional data to support the nonrisk contention. The EPA also indicated that an experimental use permit would be required prior to field-testing.

On February 1, 1985, the EPA outlined the data it required in support of the

EUP, which included the identity, detection methods, colonization potential, and competitiveness of the INA⁻ deletion mutants. The EPA stated that AGS had the burden of producing scientific data that indicated "no significant adverse effects." AGS submitted its EUP application in July 1985. The application was reviewed by the EPA's Hazard Evaluation Division (HED) and its external Scientific Advisory Panel. All the data for the application were supplied by AGS. As required under its pesticide registration rules, the EPA published an announcement of the proposal and provided a 30-day comment period. But the details of the proposal were kept confidential in compliance with rules for protecting commercial data. In response to the question of balancing proprietary information with the public's right to know, Steven Schatzow of the EPA commented:

When you have so much public concern and desire to have public education and to inform the public in this area, how can you do it when a company says, "Everything I have given you is confidential business information." All you can do is put a notice in the *Federal Register* that says, "We are having a meeting. It will be a closed meeting. We can't tell you what it's about, but we are having it."[29]

The HED was responsible for the risk assessment of the ice minus application.[30] The HED identified two types of potential adverse effects for the AGS test: the effects of ice minus on the survival and geographical range of certain plants and insects; and the effects of ice minus on climatological patterns.

For either of these effects to take place, the HED maintained that the INA deletion mutants would have to displace the normal INA population on numerous plant species over a wide geographic area. According to the HED evaluation, such displacement presupposes the occurrence of four events.

1. Once applied to the field test site, INA⁻ mutants must be disseminated outside the area.
2. Once outside the area, the mutants must survive and replicate.
3. Once they survive and replicate, INA⁻ must outcompete and displace the indigenous INA⁺ populations.
4. Finally, the INA⁻ mutants must colonize a wide variety of plants to the exclusion of INA⁺ populations.

For any organism to exhibit an effect outside the test site, conditions (1) and (2) are necessary. If the organism fails to replicate then its population will die off.

The issue of competition in point (3) deserves some elaboration. INA⁺ and INA⁻ are very closely related. They may differ only by the deletion of a few genes. Therefore, it is reasonable to assume that these organisms compete for the same ecological niche. If the niche is filled by INA⁺ strains, then INA⁻ strains will have nothing to occupy and subsequently die out. Thus, even though a microbe can survive and replicate outside of the controlled field conditions, unless there is a niche for it to occupy, a result only possible if INA$^+$ is displaced, there can be no residency and therefore no adverse consequences.

The fourth condition cited by the EPA is curious. Why is it a necessary condition for the appearance of adverse effects that INA$^-$ must colonize a wide variety of plants (outcompete INA$^+$ in many niches)? All it takes is one well-placed niche with a requisite set of circumstances for an undesirable outcome. The EPA's analysis presupposes that INA$^-$ will not occupy an entirely new niche in the ecosystem.

The HED argued that events (1) and (2) are expected to occur. Even a buffer zone around the field plot cannot prevent the test organisms from escaping and establishing themselves in some ecosystem. The major question for the EPA was whether conditions (3) and (4) would take place. Would the INA$^-$ deletion mutants displace the original INA$^+$ populations that have played a significant role in ice crystallization for plants and other media?

AGS provided the EPA with data to demonstrate that INA$^-$ deletion mutants do not have a competitive advantage over indigenous INA$^+$ bacteria when the former are applied under a variety of conditions. The only exception, according to AGS, occurs when the ice minus deletion mutants are applied at much higher dosages than the ice plus strains. For obvious reasons, if ice minus could not displace ice plus under some set of conditions, it would not be effective. AGS argued that INA$^-$ could displace INA$^+$ in the test site when the dosages of the former are high, but displacement of the indigenous strain would not take place outside the test site.

The EPA's HED drew two conclusions from the data: (1) INA$^-$ deletion mutants applied to a 0.2 acre plot will escape the plot in low numbers, but will never exceed in numbers the indigenous strains of INA$^+$ and INA$^-$ already colonized on the plant. (2) INA$^-$ will not have a competitive advantage over the indigenous bacteria on plants outside the test site. The HED therefore concluded that the proposed small-scale applications of INA$^-$ posed no significant risk of adverse environmental impacts.

To appreciate the limits of this analysis, it is instructive to recognize that the EPA's worst-case scenario assumes the 0.2 acre field site. The evaluation did not address the conditions of a widespread dispersal of INA$^-$ throughout the region.

Anticipating a legal challenge, the EPA took special precautions in setting up the review process. The HED received comments from the public, an EPA working group, USDA, NIH, FDA, and a scientific advisory panel. An EPA six-member Scientific Advisory Panel consisted of two microbiologists, a meteorologist, a community ecologist, a microbial ecologist, and a plant pathologist. The various advisory groups were in concordance with the HED conclusion that the proposed application of INA deletion mutants on 0.2 acres of land in Monterey, California, presented "no foreseeable significant risk to human health or the environment."

During the public comment period, the EPA received critical responses to the test from only one group, the Foundation on Economic Trends. Citing the impacts INA⁻ might have on precipitation patterns, the FET asked that tests be conducted to measure the upward flux and dispersal of epiphytic (plant inhabiting) bacteria.

In response to these demands, the EPA indicated that "numerous tests of this nature have been conducted to determine upward flux of *P. syringae* from agricultural crops."[31] However, the EPA did not report the results of these tests. It did indicate the possibility of conducting some monitoring experiments of upward flux—a recommendation supported by its advisory subpanel. The HED scientists consulted with several atmospheric scientists and were satisfied that the test would not affect precipitation. The precipitation problem, however, did not disappear. While, the EPA was in the throes of its review of ice minus, a letter appeared in *Science* highlighting a worst-case outcome for the release that gave credence to the precipitation scenario. The letter was written by University of Georgia ecologist Eugene Odum who, two years earlier, had joined other scientists in support of Rifkin's lawsuit on deliberate release experiments.

In his letter, Odum invoked the expertise of another scientist at the National Oceanic and Atmospheric Administration Laboratory in Boulder, Colorado, who asserted that *Pseudomonas syringae* enhances rainfall.

It seems that lipoprotein coats of this and other species of bacteria found on plants and in detritus when shed and wafted up in the clouds form ideal nuclei for ice formation that is absolutely necessary for rain to fall. . . . If *Pseudomonas syringae* does indeed have a beneficial role in enhancing rainfall, then the ecologist's concern about possible secondary or indirect effects of releases of genetically altered organisms is vindicated— incredibly, at the very first major controversy over release of engineered organisms.[32]

Odum called for a holistic approach to evaluating GEMs prior to their release. He warned against securing short-term gains in exchange for tinkering with the earth's life support systems. He used terms that have become synonymous with environmental values, such as open complex systems, holistic assessment, and the biosphere as a biogenerative and self-regulating system. This language is in stark contrast to the mechanistic and reductionist concepts in molecular chemistry. The deliberate release experiments brought these paradigms into direct confrontation (see Chapter 8).

In 1983, when Odum filed a sworn affidavit in support of Rifkin's lawsuit, he outlined the elements of a research program to be executed prior to release of a GEM. First, he wrote, there ought to be an intensive study of the organism over at least one annual cycle involving experimental releases in greenhouses that simulate as closely as possible the environment designated for the release. Second, he asked that the probability of adverse effects be examined through computer modeling in which "all conceivable situations are simulated."

The role of ice plus in precipitation was not discounted by proponents of field tests. As previously indicated, Lindow discussed this possible relationship of

INA$^+$ in his 1983 essay.[33] Odum's language expresses a more confident view about the relationship than Lindow, who characterized the role of INA bacteria in climatology as "largely unexplored." For Lindow and AGS, not only were the causal relationships speculative, but they could not foresee any circumstances under which dispersed INA$^-$ as described in the field experiments might eventually populate a region, displace their INA$^+$ cousins, and create havoc on the climate.

Three scientists at AGS rebutted Odum's published letter in *Science* about the effect of ice minus on precipitation.[34] The authors took umbrage with Odum for implying that there is proof of the role played by bacterial ice nuclei in initiating precipitation. It is a theory, they claim, that has not been tested. The conjectures emanating from the theory should not be treated as doctrine. There are two things known about INA bacteria and precipitation according to the AGS scientists: living plants, plant debris, and soil are sources of ice nucleation-active bacteria; and these INA bacteria become airborne in small numbers under natural conditions. However, they reported, there are no published data "on the concentration or activity of bacterial or bacterially derived ice nuclei in clouds."

The particular relationship between INA$^+$ and precipitation is not the key question for the risk assessment, from the AGS perspective. Instead, the crucial issue is whether "the release of INA$^-$ bacteria could have a significant impact on the number of INA bacteria available for such natural "cloud seeding." They reiterated the conclusion of the EPA that for the experimental releases contemplated, "INA$^-$ have no preemptive competitive advantage over INA bacteria."

Preparations to field-test genetically modified organisms at designated sites intensified coverage of these debates in the media. During 1986, there were three communities slated for such field experiments, two in California for ice minus bacteria and one in Missouri for a naturally occurring microbe into which a *Bacillus thuringiensis* toxin gene was inserted. Ice minus proved to be the path-breaking experiment that challenged the regulatory process and ushered in a new era for biotechnology.

COMMUNITY RESPONSE TO ICE MINUS FIELD TESTS

The development of an effective strain of ice minus was being pursued by scientists at UC Berkeley and AGS. Both institutions were preparing to field-test their respective organisms. The Berkeley scientists were first to request and receive approval from NIH to release ice minus into the environment in a controlled field test. The FET lawsuit against NIH forced a postponement of the scheduled test date of May 1984.

Advanced Genetic Sciences submitted its field test proposal to the RAC in March 1984. By October, the NIH notified the company that its field experiment would be approved pending further data and a monitoring plan. However, since the EPA was advancing a new regulatory regime for biotechnology, in November AGS inactivated its application to the NIH and submitted a notification of a field test to the EPA's Office of Pesticides and Toxic Substances. After a review period of 13 months, AGS met the EPA's requirements for an Experimental Use Permit (EUP).

The EUP issued to AGS was for field-testing ice minus at a site (as yet undisclosed to the public) in Monterey County, California. A local Christmas tree farmer who read about the unprecedented test gathered a few signatures and registered a complaint to the county board of supervisors. The letter of complaint cited the secrecy of the test site and the unknown risks associated with the organism. The technical detail cited in the letter reveals that the Monterey petitioners were in prior contact with informed sources.

We are concerned about the planned test involving the genetically altered bacteria, Frostban, by Advanced Genetic Scientists [sic] in north Monterey County. Numerous questions about the test remained unanswered. A primary concern is that there are no guarantees that the bacteria will not spread beyond its designated test site. A fifty to sixty foot buffer zone is not sufficient to prevent the bacteria from being blown or moved by other natural phenomenon into non-designated areas. The possible side effects of Frostban on the environment are unknown factors that could affect rainfall patterns since the organism in its natural state is an important agent in cloud formation. Furthermore, few studies on the toxicity of Frostban have been completed. Thus, the possible impact on local crops remains unknown. Another concern is that the local authorities do not need to be notified of the whereabouts of Frostban's application until twenty-four hours before its use.[35]

A telegram to the board that was highly critical of the proposed field test came from the Green Party members of the West German parliament, who warned of "grave dangers to the environment" and cast the test as an international event.

It is not only the citizens of Monterey County who will be affected if this irreversible experiment is allowed to proceed, but the citizens of our own country as well, and indeed of all nations. Our health and environment must not be sacrificed to the commercial interests who are so eager to bring their new living and reproducing products to the marketplace. . . . We urge you to use your powers to prevent this potentially dangerous experiment from taking place in Monterey County and to open the door instead to greater public scrutiny of the crucial issues involved by authorizing a public hearing. Here in the Federal Republic of Germany we currently have a regulation which prohibits any release into the environment of a genetically-engineered organism, and we are striving to keep it in effect.[36]

A second telegram was signed by Green members of the European Parliament situated at the time in Strasbourg, France: "After all scientific information available about this genetically engineered bacteria, which is at the present state of art by far insufficient, the deliberate release of 'ice-minus' would be a grave, yet maybe irreversible hazard to the environment."[37]

The Monterey Board of Supervisors held the first of a series of hearings on January 27 and declared a 45-day moratorium for such field tests on February 11.

Monterey citizenry became more incensed over the field experiment when it was disclosed that the site selected was adjacent to local residences. To make matters worse, a disgruntled employee of AGS blew the whistle on the company by disclosing that ice minus had been injected into trees located on the roof of the facility without EPA approval. A subsequent EPA investigation charged that AGS falsified information on the EUP application (a charge later dropped) and failed to perform tests according to agency requirements (a charge that held up).

The accumulation of allegations and the company's ineptness in communicating with local authorities severely tarnished AGS's credibility within Monterey County. The EPA withdrew the EUP for the field test. And nearly a year and a half after the first test was planned, the Monterey Board of Supervisors passed an ordinance regulating the release of GEOs based upon the county's authority over land use.

Following a parallel track, UC Berkeley scientists Lindow and Panopoulos received EPA approval to field-test their ice minus strain at the state agricultural field station in the northern California community of Tulelake. Two women from the town began an organizing effort to stop the test after they had learned about ice minus from a Monterey citizen activist who addressed the Tulelake community. The Tulelake organizers built a small grass-roots committee on the model of the Monterey activists, calling themselves the Concerned Citizens of Tulelake (CCT). They notified local governments in contiguous communities, drew up a petition, lobbied various groups, sent out a stream of irate and informed letters to federal, state, and local authorities, and organized their own local forum on the release of "mutant bacteria."[38] The CCT also played a significant role in shaping local news coverage about ice minus.

By June 1986, the CCT was successful in convincing both the Modoc and Siskiyou County Boards of Supervisors to pass resolutions opposing the ice minus experiment. Undaunted, UC Berkeley continued its plans to test ice minus, threatening legal action against local or county opposition.

The lawsuit came instead from the CCT, with help from Jeremy Rifkin and Californians for Responsible Toxics Management (CRTM). The court issued a restraining order blocking UC Berkeley from undertaking the field test. Within several weeks, the parties to the suit reached a negotiated settlement under which the University of California was responsible for issuing an environmental impact review (EIR) and in return the citizens groups would withdraw their petition.

While UC was preparing its EIR, AGS decided not to pursue a field test in Monterey County but proceeded to enter into negotiations with other California communities. This time the company took a proactive role in communicating directly with farmers and civic organizations. It declared that it would not pursue a site in secrecy or under conditions where community approval was not forthcoming.

AGS finally met the EPA requirements for the field test in the fall of 1986. By February of the following year, AGS had endorsements for its tests from two county boards of supervisors and approval from the EPA of three sites. An

132 Genetics and Ecology

environmental coalition including Earth First, the Berkeley Greens, and FET failed in a final effort to gain a court injunction against the test planned for Brentwood, California. The first federally sanctioned deliberate release of genetically modified bacteria took place amidst throngs of reporters on April 24, 1987. Five days later the UC Berkeley test was performed without the attention of the national media.

From the time the two Berkeley scientists submitted the field test proposal to the NIH, it took nearly five years for the legal, political, and regulatory requirements to be satisfied. Because it was first, ice minus carried a lot of weight. It symbolized a new era for biotechnology, the transition from the laboratory to the field.

Despite the attention it received, there was broad consensus among scientists that ice minus was probably among the safest agricultural products one could create with the new technology. The organism had genes deleted and not added, and there was no evidence that the product could outcompete its parental strain. As a public controversy, ice minus was history just as soon as the field was sprayed by the moon-suited scientist. A precedent had been set insofar as the first major barrier to the environmental application of GEMs was overcome.

Two issues remained. Would ice minus become a viable product? And, can the scientific community reach some consensus on evaluating the risks of other deliberate releases? Ice minus was the catalyst that fostered a dialogue between two communities in science that until this time had been well insulated from one another. The next chapter explores how different segments of the scientific community assessed the risks and approached the problems of environmental releases.

8

Debates over Deliberate Release:
Disciplinary Fault Lines

We are not Luddites or alarmists, but merely skeptics who wish to consider
what the hidden costs of this promising new technology might be.
                                                    Frances E. Sharples, 1987[1]

In trying to assess the potential dangers, the experience of ecologists with
transplanted higher organisms is less pertinent than are the insights of fields
closer to the specific properties of engineered microorganisms.
                                                    Bernard Davis, 1987[2]

In the development of science, two types of disorders or crisis may be distinguished. Disorders of the first kind refer to the crises that arise over scientific
theories or what Thomas Kuhn has described as paradigm conflicts.[3] A theory
confronted by anomalous data or two theories competing for hegemony in a
divided discipline are examples. Sometimes a critical thought experiment may
elicit a contradiction in a theory that places its credibility in question. Other
times the theory may be faltering under new experimental evidence. Disorders
of the first kind originate from within science and often bring critical attention
to foundational issues. Resolutions of the crisis may involve tinkering with
theories, paradigm shifts, or a full-fledged scientific revolution.

Disorders of the second kind surface when science is outer-directed, that is
when theory is used to solve problems arising out of a social context. An
important class of problems in this category are the applications of science for
the assessment of technological risk. Scientists have been divided on the risks
of chemical carcinogens, nuclear power, ozone depletion, and carbon dioxide
buildup, along with many other outcomes of modern technology.

In 1975, disorders of the second kind were highlighted in the debates over
the potential hazards of research involving recombinant DNA molecules. Molecular geneticists turned scientific theory on itself, namely, to study the risks

that the theory helped to create. Within a few years the boundary of disciplinary participation in the rDNA debate expanded to include both infectious disease epidemiologists and immunologists, who contributed their perspectives on the risks of gene-splicing experiments.

When science is outer-directed and applied to public policies, disagreements may arise over methodological assumptions, interpretation of theory, or the relevancy of theoretical principles to practical problems. The nature of credible fact is often in dispute. Abstract models take the place of replicable experiments. The term "second order" science was introduced by S. O. Funtowicz and J. R. Ravetz to describe the application of scientific approaches in the face of high uncertainty.

Science cannot always provide well-founded theories based on experiments for explanation and prediction; but can frequently achieve at best only mathematical models and computer simulations, which are essentially untestable. On the basis of such uncertain inputs, decisions must be made under conditions of some urgency. Therefore science cannot proceed on the basis of factual predictions, but only on policy forecasts. Typically, in such cases the facts are uncertain, values in dispute, stakes high and decisions urgent. In this way it is "soft" scientific information which serves as inputs to the "hard" policy decisions in many important environmental issues.[4]

Disorders of the first and second kind are similar in the importance that the relevant parties place on disciplinary autonomy and competition over a preferred paradigm. No disciplinary group likes to give up control over a subject about which it believes it has standing. Each discipline prefers to see its construction of reality and its claim to truth predominate.

In the early rDNA debates, molecular geneticists struggled to retain control over how the risks were defined and safety guidelines were written and imple-mented.[5] A decade later, a similar struggle has been taking place between ge-neticists and ecologists regarding the deliberate release of GEOs. In this case the disciplinary forces are more evenly divided and sharply contrasted. The issues are no longer laboratory-based since industrial applications of novel organisms are being considered.

I have chosen to examine the scientific uncertainties over deliberate release because it offers an opportunity to raise epistemological questions about paradigm construction and disciplinary hegemony in the context of a public policy debate. Molecular geneticists created gene-splicing and made preparations to test new plants and microorganisms in open fields. They were impelled by guidelines and regulations to evaluate the risks. As progenitors of the technology and those with continued interest in its development, they wished to maintain control over its use, and therefore over its risk assessment. The prospect that experiments might be restricted, delayed, or even banned represented a crisis to those working in the field of applied genetics.

Ecologists define their professional domain to include ecosystems comprised of plants, insects, and animals. When regulatory authority for deliberate release shifted from the NIH to the EPA, ecologists began playing a stronger role in the debates over safety.

The stakes were different for geneticists and ecologists. While the former wished to maintain control over the use of the technology, the latter sought disciplinary standing and influence over the standards of technology acceptance. Moreover, they viewed the issues related to deliberate release in terms of greater funding for ecological research.

Molecular genetics is a relatively new and fashionable field of biology. While its roots go back to the 1940s with the pathbreaking work of Max Delbruck and Salvador Luria,[6] the field exploded in the mid-1970s. Molecular biology has been attracting substantial funding both for basic and applied research in medical and agricultural areas. The new breed of genetic alchemists have been acquiring a sizeable share of Nobel Prizes as well as other professional accolades.

Ecologists, on the other hand, could not make a claim to either the financial or professional stature in the scientific community that molecular genetics had attained. But they could make claim to the study of natural ecological systems and the introduction of exotic species into those systems.

This chapter explores how the risks of deliberate release experiments are addressed by two centers of disciplinary activity. I shall argue that there is a strong relationship between the scientific paradigms employed and the approach taken to risk. The theoretical program of geneticists emphasizes nature's unity and stability, and expresses confidence in human control over biological systems.

In contrast, the perspective of ecologists focuses on nature's complexity and interdependence. Ecologists are more comfortable studying nature as a dynamical system comprised of nonlinear processes that give expression to nature's indeterminacy.

Crisis in science, whether of the first or second kind, can thrust it into higher stages of self-examination and self-criticism. In this case we see competition between two disciplinary centers, each vying for dominance in the public policy arena. The focus of this policy debate is the introduction of genetically modified organisms into the environment.

ANTICIPATIONS OF DELIBERATE RELEASE

Plans to release genetically engineered microorganisms into the environment were widely publicized after the 1980 Supreme Court decision on the patenting of life. The court ruled favorably on a patent for an oil-degrading organism that was developed by Ananda Chakrabarty who, at the time, was working for General Electric. The new bacterium was designed to be spread over oil spills. Offshore oil drilling had intensified along the coastline of the United States in the aftermath of the 1973 oil crisis as new government incentives improved the economics of exploration. The massive oil spill off the coast of Santa Barbara in the late 1960s showed how vulnerable an ecologically sensitive coastline was to a major accident.

The Chakrabarty organism was not constructed by recombinant DNA techniques. Plasmids from separate organisms were placed into a single bacterium. Each of the plasmids has the ability to degrade an important component of the hydrocarbons comprising crude oil. By having the plasmids in a single organism, it was believed that the four major components of crude oil could be efficiently broken down into harmless products.[7]

Much of the discussion around the oil-degrading bacterium was centered on legal and ethical issues such as "should one own the blueprint of life?"[8] There were, however, a few faint voices that questioned the ecological impacts of releasing oil-degrading organisms into the environment. They postulated scenarios in which the organism might run amok and degrade useful hydrocarbons. However, no public record was established of any serious effort to evaluate the ecological consequences of releasing the novel organism.

Despite the favorable patent ruling by the Supreme Court and its boost to the fledgling biotechnology industry, General Electric did not develop and market the oil-eating organism.

Several explanations have been advanced to explain the fate of the GE microbe. First, there is the pure technological hypothesis, namely, that the microbe was not efficient for large-scale hydrocarbon degradation. The success of the product may be limited to small-scale, highly controlled conditions.

Second, an economic hypothesis holds that the product was not financially viable. The number of oil spills resulting from deep sea drilling had diminished, probably as a result of two factors: (1) there was less American oil exploration in the aftermath of dramatic reductions in world oil prices by the late 1970s; and (2) the safety of oil drilling improved as a consequence of major spills like the one off the Santa Barbara coast.

According to a third hypothesis, the organism was likely to meet too much public resistance, particularly at a time when liability insurance for deliberate releases was uncertain and there were no regulations in place.

Despite its fate, the product was an important prototype for a new generation of biodegradative organisms. With tens of thousands of chemical waste sites in the United States and countless thousands throughout the world, microorganisms genetically engineered to degrade chemicals like PCBs and dioxins that have contaminated soil could be in great demand. A major Alaskan oil spill in 1989 brought a renewed interest in a microbial process for responding to oil tanker leaks.

In 1980, the experimental oil-eating microbe was not a subject of discussion among ecologists, possibly because the product was not being seriously considered for environmental release. But this doesn't fully explain their lack of involvement since proposals to release genetically modified plants also did not draw critical attention. It was the ice minus proposal first introduced in 1982 that brought ecologists face to face with molecular geneticists on evaluating the risks of releasing genetically engineered microorganisms (GEMs).

It is far too simplistic to say that the debate over deliberate release sharply divided ecologists and molecular geneticists exclusively. As indicated by Peter Day, the shadings and variations among the disciplines are far more complex than is suggested by a simple bifurcation.

Ecologists, environmental scientists, and agronomists are concerned with populations of organisms and how they behave. Plant breeders, microbiologists, and taxonomists work largely with single organisms either as individuals or as large but uniform populations. Cell biologists, physiologists and many biochemists work with parts of organisms, either single cells or homogeneous tissues. Molecular biologists work at the subcellular level with organelles and molecules.[9]

Each discipline and subdiscipline is serviced by its unique vernacular and conceptual architecture through which it describes and interprets some component of nature. It should not come as a surprise that representatives of the respective disciplines would approach the problems of risk differently. The dialogue between molecular geneticists and ecologists has focused the debate around certain principles about organisms and natural systems that lie at the root of the risk assessment enterprise. Policy issues related to the applications of science and technology can at times stimulate controversy about scientific principles and their interpretation. This is particularly evident in the area of risk assessment, where policy rests on shaky, tentative, or trans-scientific judgments. To illustrate this I have selected a few issues that draw attention to the differences between organismic and molecular approaches to biology. It is a conflict of traditions and paradigms that is reminiscent of the eighteenth-century debates between vitalists and materialists. Philip Regal, ecologist and behavioral biologist at the University of Minnesota, highlights this point:

We see a disagreement, an inevitable stand-off between one group of scientists whose studies have for nearly 25 years been revising their views of design and simplicity of nature, and some devotees of another tradition that has often taken it on faith (and on success in their own limited domain) for over 50 years that nature is simple and that "true" science begins with this assumption of essential simplicity.[10]

In a similar way, Harvard biologist Ruth Hubbard frames what she believes is the critical parting of the roads in modern biology.

Nothing illustrates the dichotomy more clearly than the controversies that have arisen over genetic engineering of bacteria, plants, and animals, and more recently over the Human Genome Initiative, the projected effort to identify the genes and establish the sequence of all the bases on all twenty-three human chromosomes. . . . All these are practical implementations of the reductionist belief that organisms are merely the living manifestations of their DNA. . . . On the other side are people who see organisms as more than the sum of their parts, who believe that connections and process cannot be understood by isolating things or events, who distrust reductionist oversimplifications.[11]

The following discussions illustrate the view that those who think reductively about their science also show a tendency to think reductively about risk. The belief that DNA is the Holy Grail for explaining and predicting all there is to know about organisms and their behavior easily translates to an understanding (or misunderstanding) of environmental risk.

NOVELTY AND EVOLUTION

One of the arguments in support of granting special attention to organisms developed by gene-splicing methods is that the resulting products are unique to nature and therefore may be capable of escaping nature's self-correcting system of controls. The counterargument is that genetic exchange and natural recombinations are constantly taking place. Moreover, the argument continues, over several billion years of evolution nature has tried variants of organisms with all possible gene combinations. In nature's experimental lottery, only a small number of the organisms had what they needed to compete. Scientists who produce a so-called novel strain are most likely recapitulating one of nature's failed experiments.

The argument that nature has tried variants of organisms with all possible gene combinations appeared in the earlier rDNA debate when the focus was on laboratory experiments. This line of reasoning was contested by Richard Novick,[12] who calculated that only a fraction of all possible genetic combinations could have been tried in the course of evolution, and more recently by Sharples,[13] who maintains that occasionally nature produces something new, such as the AIDS virus.

As regards the potential hazards of deliberate releases, it hardly matters whether or not an organism has had exposure to certain foreign genes at some distant point in its evolutionary history. The environments on earth have varied significantly over long evolutionary periods. Since the ability of an organism to survive is so linked to the environment, for all practical purposes, an organism with foreign genes introduced 100 million years ago can tell us almost nothing about how the same organism will behave today if released into the environment.

FITNESS AND FOREIGN GENES

What happens to the fitness of an organism when a few foreign genes are inserted? Is it likely to survive and spread into the environment or will the added genes be a liability to survival? Can a few foreign genes radically change a nonpathogen to a pathogen? The answers to these questions are clearly germane to the problem of hazard assessment. If the addition of foreign genes makes a microorganism less robust than its parent strain, an important margin of safety is built into the genetic engineering process. Plant geneticist Winston Brill developed this argument in an essay published in the journal *Science*. According to Brill, since the organism has not evolved with the added genes, "the extra burden to the organism carrying new genes should decrease its ability to compete and persist."[14]

The view that changing an organism by adding new genes makes it less competitive has a theoretical and an empirical basis. On the theoretical side, it is argued that organisms, including their genome, are highly perfected or optimized to their environment through millions of years of evolution. Any change to any part of the organism will reduce its optimality. Applying an organismic perspective, Washington University geneticist Daniel Hartl frames the argument as follows:

Organisms represent more than the sum of their parts. Genes interact with one another in complex primary and secondary ways, and a living organism represents a successful integration of those manifold interactions. Thus, the most likely outcome of the introduction of alien genetic material into an organism would be the disruption of this integration. Far from becoming a superior competitor, the organism would be likely to become hampered in its ability to compete.[15]

The empirical side to the fitness argument draws upon the experience of conventional genetic alteration. For decades prior to the discovery of gene-splicing, scientists were mutating organisms with chemicals and radiation and observing their effects. According to Hartl, "most mutations that result in visible effects on phenotype are harmful to the organism, usually very harmful . . . successful organisms result from favorable interactions of genes among themselves and with the environment. Major changes in phenotype are likely to impair interactions, not improve them."[16]

A variant of the previous argument described by Martin Alexander maintains that by adding new DNA to an organism, it will not be able to compete with organisms that do not have the excess baggage. Alexander acknowledges an opposing view, however, which states that "if genetic engineering can confer on any organism a trait that makes it a better competitor, or resistant to toxins, predators or parasites, the organism might not suffer appreciably even if it had to expend more energy to synthesize additional DNA or enzymes."[17]

Even if the preceding conclusions are generally correct, the few exceptions to the rule are what concern some ecologists. All it takes, so to speak, is one statistical anomaly to create an environmental problem. Philip Regal takes exception to the argument that species are "exquisitely modified for particular functions in nature" and that "any artificial modification of a natural form will cripple it." The term "cripple" may be an overstatement in this circumstance, since geneticists like Brill and Hartl suggest that added genes give the organism a "relative disadvantage." According to Regal, "when some genetic engineers insist that adding creatures with new biological properties to the environment will always be safe, this at once raises the spectre of the long history of costly and tragic biological misjudgements."[18]

Regal drives a wedge between the biochemist/geneticist and the organismic biologist on genetic fitness and its role in risk assessment. Referring to the optimum fitness position, Regal stations himself firmly on the evolutionary record: "This line of reasoning is based on ideas that have quite a long and infamous history in ecology, but have been driven into exile in the eyes of those ecologists and evolutionary biologists who have faced them squarely."[19]

The dispute hinges on the distinction between adequate versus optimal fitness of evolved organisms. If every evolved species were optimal to its environment, then any alteration in the genome would make the organism suboptimal. Regal disputes the optimality hypothesis and contends that organisms are adequately, not ideally, constructed for survival. "Organisms are not perched on narrow adaptive peaks."[20]

Regal does not dispute the probability that introducing foreign genes into an organism reduces fitness, but he does argue against a direct causal relationship. "There is no evidence that every artificial modification by recombinant DNA techniques will necessarily attenuate its ability to function in nature."[21] Support for this position is given by ecologist Frances Sharples, who served on the NIH's Recombinant DNA Advisory Committee. Sharples courts evolutionary theory to defend her thesis.

The fundamental premise of evolutionary theory is that natural selection, the dominant force responsible for adaptations of organisms to their environments, operates on genetic alterations or novelties—mutations, rearrangements, and acquired accessory elements, such as plasmids—to produce evolutionary change. It follows that at least some genetic alterations improve the abilities of organisms to survive, reproduce, compete for resources, or invade new habitats. . . . A general assertion that genetic alterations, be they natural or man-made, always lower the fitness of organisms is therefore not warranted and runs counter to basic evolutionary principles.[22]

The fitness argument has been used since the early days of the rDNA controversy to fend off regulation. Those opposed to regulation argued: Why worry about an organism doing harm to humans or the environment if it will not survive? But the point of releasing a genetically altered strain is based upon the presumption that it *will* survive at least as long as is necessary to carry out its function. Organisms designed for use in the environment must have some survival capacity. To be effective, Chakrabarty's oil-degrading organism must survive long enough to consume the crude oil. Ideally, once that job was done, it would self-destruct.

For the *Pseudomonas* bacterium with the deleted ice-nucleation gene to be effective, not only should it survive, but it must displace the indigenous strain (ice plus). Similarly, if plants are designed to fix nitrogen or harbor a toxin to control pests, they are being engineered *for survival*. If the organism cannot compete for survival then, from the commercial standpoint, either the environmental niche designated for the biological entity must be modified to improve its fitness or the entity will not prove useful. Consequently, the fitness argument may be moot for environmental applications of genetic technology. An organism pursued for environmental purposes will be designed to survive for some period in some niches. The next obvious question is: Under what circumstances will a GEO exhibit unexpected (in relation to its parental strain) adverse properties?

rDNA AND UNIQUE HAZARDS

In 1977 the National Academy of Sciences (NAS) sponsored an Academy Forum on recombinant DNA research. The agenda included panels of scientists, paired in adversary fashion, to address the anticipated applications and liabilities of the new research technology. It was at this forum that Jeremy Rifkin launched his public advocacy campaign against genetic engineering in the form of a "guerilla theater" protest.[23]

Ten years later, the academy had its second major public relations encounter with genetic engineering with the publication of a pamphlet titled *Introduction of Recombinant DNA-Engineered Organisms into the Environment: Key Issues*.[24] The document represented the consensus position of five scientists; in addition it had the imprimatur of the NAS. The pamphlet is noteworthy for its brevity, simple language, and forthright conclusions about the scientific basis for releasing genetically modified organisms in the environment.

In response to the question "can we know?" the report states affirmatively: "Adequate scientific knowledge exists to guide the safe and prudent use of R-DNA-engineered organisms in the environment and to identify the most problematic introductions."[25] This conclusion runs counter to the belief of leading ecologists that predictive knowledge about safe releases is still in its infancy and current methods of evaluating risks are unreliable.

Among the most significant conclusions in the report is a statement about "unique hazards." Since the discovery of rDNA techniques, scientists have debated whether their use would result in hazardous products that were unique to this technology. Early claims by scientists that rDNA techniques could "breach species barriers" provided grist for the uniqueness hypothesis.[26] Failing to find support for the supposition, the academy report rejected the view that rDNA technology introduced qualitatively new risks: "There is no evidence that unique hazards exist either in the use of R-DNA techniques or in the movement of genes between unrelated organisms."[27]

This statement makes its appearance as a sober scientific judgment. Actually, it is embroidered on a background of tacit and unreflective assumptions. To illustrate this, we simply inquire: What evidence would support the thesis that rDNA techniques are associated with unique hazards? Until we understand the form that positive evidence would take, we cannot appreciate the significance of the negative claim.

In order to demonstrate that rDNA techniques can produce unique hazards by the academy's criteria we have to show two things: (1) there is at least one hazard that results from the application of the techniques; (2) the hazard is the result of the techniques exclusively (i.e., it cannot arise from conventional genetic exchanges or natural exchanges).

Proposition (1) has not been validated. No evidence of a hazard resulting specifically from rDNA techniques has been demonstrated beyond the hazards associated with the starting materials. In the case of deliberate releases, there have been very few field tests from which to draw any generalizations.

For the moment, let us assume that condition (1) can be met (e.g., an organism is given an added phenotype such that, when released, it adversely affects a segment of the ecosystem). To meet the conditions of uniqueness we still must demonstrate that conventional methods of DNA transfer cannot be used to produce a similar result.

While proposition (1) is widely (but not universally) accepted, the authors of the NAS position paper offer neither criteria nor empirical data that could be used to support proposition (2). Unless we understand what counts as supporting data or evidence for this second condition, it makes little sense (except perhaps for rhetorical reasons) to speak about unique hazards.

How can one prove that an rDNA-derived hazard cannot be produced by one of several conventional techniques for transplanting genes, a necessary requirement for demonstrating uniqueness? One would have to make a bold effort to construct the hazard by techniques other than rDNA. Only when the hypothesis stands up against efforts at refutation can we have confidence that there are unique hazards in the rDNA process.[28]

In the final analysis, the NAS claim about the nonexistence of unique hazards has the flavor of ideology and not science, since those who dismiss the hypothesis do not define the conditions of its verifiability. There is one sense in which the hypothesis is trivially true. Since there is no evidence of a hazard, there is no evidence of a unique hazard. But the debate is not about what our scant empirical data reveals to date. It is about the limits of the possible. Therefore, whether or not rDNA is capable of unique hazards, the issue over ecological risks continues to pose a challenge.

GENETIC PREDICTION AND ECOLOGICAL HAZARDS

What is the state of prediction in molecular genetics and ecology? By modifying a few genes in an organism, what is the likelihood it will behave unexpectedly in the ecosystem? Where do scientists derive their predictive confidence: the genetics side or the ecological side?

These questions were at the epicenter of a spirited debate spurred by various initiatives to regulate biotechnology. One argument vigorously promoted by

Frank Young, former member of the RAC and commissioner of the Food and Drug Administration, and his deputy Henry Miller holds that rDNA techniques provide far more precise and predictable changes in an organism than any other genetic techniques. Why, they ask, should this technology be selected out for special regulation?[29]

There is an important sense in which the authors are correct. Compared to gene-transfer techniques that shotgun undefined segments of DNA into a microbe or compared to hybridization techniques, the transfer of well-characterized DNA segments is more precise and more predictable. But, for purposes of hazard assessment and regulation, the comparison between conventional and modern genetic techniques is misleading. Conventional genetics, after all, has not been the springboard for the type of industrialization that has resulted from modern biotechnics. Regulations were directed at rDNA techniques precisely because they opened up the floodgates for the commercialization of genetically modified organisms.

The essential question is not whether we can make better predictions with rDNA techniques over those of conventional genetic processes, but how well we can predict the environmental effects of GEOs, since rDNA is the preferred genetic exchange process for modern biotechnology.

A response to the claim that precise genetic alterations secure safer outcomes was addressed in a report prepared by seven scientists for the Ecological Society of America (ESA) in 1989. The ESA report cited cases of secondary phenotypic effects resulting from a single genetic alteration. Some of these effects are only expressed in certain environments. The ecologists cite the importance of examining the characteristics of the GEO (genes and phenotype) as well as the environment for which it is being designed.[30]

The changes of interest in deliberate release experiments are the transformation of a plant into a weed, or changing a nonpathogen into a pathogen. A notable exchange of letters took place in *Science* during the summer of 1985 in response to a published essay in that journal by plant geneticist Winston Brill. Brill's article was titled "Safety Concerns and Genetic Engineering in Agriculture."[31]

In his essay, Brill argued that genetic exchanges are not as unpredictable as some have made them out to be. For example, the chance of creating a weed by crossing two nonweeds is exceedingly small. In some important respects, Brill contended, the process of synthesizing chemicals introduces more uncertainty than that of genetically altering organisms. The claim turns the old vitalist debate on its head.

Even if a new chemical is only a slightly modified analog of a known safe chemical, the degree of safety cannot be extrapolated from the safe chemical. In fact, analogs of normal metabolites can be most dangerous. By comparison, minor modifications obtained by breeding safe plants or mutating safe microbes do not yield progeny that become serious problems. Minor modifications are expected from genetic engineering. A program that aims to utilize, in agriculture, a plant, bacterium or fungus considered to be safe but with several foreign genes will have essentially no chance of accidentally producing an organism that would create an out-of-control problem.[32]

Brill's views on the improbability of radically altering the phenotypes of a biological entity by genetically modifying a small percentage of the genome were echoed by other geneticists. Former RAC member Nina Federoff commented: "We have a great deal of knowledge that allows us to predict what a genetically engineered organism will be like compared with its parent organism."[33] British plant geneticist Peter Day remarked that "nothing has yet happened to tell us that an organism with recombinant DNA is any more dangerous than the sum of its parts of which it is composed."[34] Waclaw Szybalski, editor-in-chief of *Gene*, concurred with a modest proviso: "Statistical laws imply that pests which cause great harm could not be inadvertently produced by genetic engineering from innocuous organisms without being preceded by an early warning consisting of the appearance of some weakly harmful constructs."[35] Szybalski's claim, like the others, is based on an idea of dubious empirical merit—that biological nature abhors discontinuous changes.

One approach favored by geneticists for evaluating the risks of gene manipulation is to examine the host organism and the foreign genes designated for implantation. Laboratory tests and field experiments play a less significant role in risk assessment than the analysis of the elemental parts in the experimental system. The striking difference between the molecular and organismic approaches to risk was highlighted in congressional testimony by Elliott Norse, director of the Public Affairs Office of the Ecological Society of America.

Molecular biologists work in laboratories to penetrate the mysteries of the invisibly tiny world within cells. In contrast, for the most part, ecologists work in the fields and wetlands and forests to unravel the mysteries of nature at a vastly larger scale. . . . Rather than working on the very smallest, most isolated components of life in situations far removed from nature, our work examines the interactions of living things in the fascinatingly complex world that we can see.[36]

When ecologists address the idea of "predictability" for genetically modified organisms, their paradigm extends beyond the organism per se to the environment into which it would be released. Since environments are far more varied and less controllable than genetic determinants, this adds a significant factor of uncertainty to risk assessment.[37]

Despite the differences in their approach, however, ecologists and geneticists are in general agreement that most species, whether non-native or genetically modified, would fail to become a significant competitor to indigenous species. But ecologists have no illusion about predicting the behavior of GEOs released into the natural environment. Martin Alexander's sentiments are typical of those of his colleagues: "Ecologists are unable to predict which introduced species will become established and which will not, and it is often not possible to explain successes or failure after the fact."[38]

Echoing this position, Cornell entomologist David Pimentel addresses the possibility of anticipating an ecological catastrophe. "The probability of an environmental disaster after a single release of any engineered organism cannot be accurately predicted at this time."[39] Philip Regal maintains that there is no predictive theory that connects the genetic sphere with the ecological sphere. "Completely satisfactory categories of *a priori* risk are not possible. . . . Genetic 'engineering' is actually largely genetic tinkering."[40]

When the Office of Technology Assessment (OTA) issued its report on field-testing genetically engineered organisms in 1988, the analysis reflected the conclusions reached by the ecological community on the lack of a predictive framework for assessing field experiments. The OTA stated that "at present, standardized protocols do not exist for predicting either the kinds of risks or the magnitudes that could be associated with planned introductions of genetically engineered organisms into the environment."[41] However, the OTA understated the complexity of the problem by speaking of "standardized protocols" at a time when ecology is without any protocols, standardized or not.

If the genetic model fails to give us a profile of the organism in the environment, and ecology is without a predictive framework, how far can ecology go toward developing such a framework? Harvard bacteriologist Bernard Davis argued that the field of ecology is by its very nature descriptive, and therefore the problem of prediction is more than a question of the theoretical maturation of the discipline. Ecologists have not been clear on the potential of their discipline to develop predictive models.

Ironically, Davis supports a risk assessment based upon theoretical principles—but principles drawn from evolutionary theory (what Regal refers to as the a priori method)—rather than empirical approaches involving elaborate and expensive field tests.

In stark contrast, a report by the Ecological Society of America states that theoretical principles and laboratory data, while both necessary, are insufficient for hazard assessment. "[Laboratory] data alone cannot accurately predict the fate of an introduced organism released in nature . . . the usefulness of data . . . for predicting environmental fate will vary widely."[42]

Thus, while ecologists acknowledge the limitations for prediction both in genetics and in their native discipline, geneticists have significantly more confidence in their predictive capacities. In response to the poverty of predictive theory, ecologists turn to empirical observations of GEOs under simulated and actual field conditions. They have proposed three stages of analysis prior to a full-scale release of a GEO.[43]

1. Intensive study of releases in phytotrons (greenhouse microcosms) that simulate the actual environment

2. Computer modeling of adverse effects in which a wide array of situations is simulated

3. Studies of the organism under actual field conditions

But these conditions represent a ponderous requirement for a regulatory system that seeks an efficient review process for the rash of new product applications. A rigorous, multi-year, case-by-case review for each application threatens the success of the new biotechnology enterprise. The "science" of risk assessment is beginning to adapt to the requirements of industrialization. Since ecologists are not yet invested in the development side of biotechnology, their adaptation has been slower.

EMERGENT PROPERTIES

In a joint letter to *Science*, Elliott Norse, joined by four nationally known ecologists, contested Winston Brill's approach to risk assessment and emphasized that his was a "geneticist's evaluation of potential ecological hazards." "As ecologists," they wrote, "our evaluation of such hazards is quite different."[44]

The authors of the letter contested two components of Brill's argument: that engineered plants are even less likely to cause problems than plant varieties produced by conventional technology; and that conventionally produced varieties have caused no problems in the past. The second proposition, they claimed, was patently false, while the first does not square with sound ecological principles. They called into question the claim that by introducing well-characterized genes into an organism, unexpected alterations will not take place. The authors cited experimental evidence that the transfer of characterized plasmids produced unexpected consequences that might influence the ecology and population dynamics of the organism.[45] According to Guenther Stotzky of New York University, when foreign genes are introduced into plasmids, sometimes new properties of the cell are observed.[46] This is attributed to regulatory upset in the cell because of the presence of the new genes. Referring to some Soviet studies on plasmid transfer, Stotzky cites examples of unpredicted or unanticipated changes in genetically engineered organisms. "The organisms were resistant to other antibiotics, their virulence was very greatly enhanced, their whole biochemical patterns changed, and they now use amino acids and organic acids that they didn't use before."[47]

The position one takes about the emergent effects of genetic engineering has an important bearing on how one views the risks. The prospect that even a few genes can radically transform an organism places a much greater burden on developing dependable tests prior to the release of an agent into the environment. Ecologists have argued that it is not sufficient to evaluate each component of the genetically constructed entity as a means of understanding the product. In the view of Harvard biologist Fakhri Bazzaz, moving genes from one organism to another changes the genetic neighborhood, and "knowing what the gene does in one organism does not necessarily mean that we will know exactly what it might do in a different organism in a different neighborhood."[48]

From that perspective, one must anticipate the unexpected and plan for it. The dictum of the ecologists has been "expect surprises." For this reason, they have called for a greater emphasis on *in situ* studies (microcosm, greenhouse, and field tests).

From a regulatory standpoint, the presumption that a genetic engineering product will not exhibit a novel property vastly simplifies the review process. Under such a presumption risk assessment is linear: from the properties of the parts, the properties of the whole can be predicted. A somewhat weaker but related presupposition is that nonpathogens do not beget pathogens and nonweeds do not beget weeds. A study by the National Research Council concludes: "It is highly unlikely that moving one or a few genes from a known pathogen to an unrelated nonpathogen will confer pathogenicity on the recipient."[49]

By questioning these presuppositions, ecologists have elevated the uncertainty of the risk assessment. For example, what criteria can be used to evaluate the unexpected in the case where foreign genes and a host organism are each considered safe? Can test protocols be developed that are likely to reveal emergent properties or secondary phenotypic effects?

SCIENCE AND REGULATION

The prospect of regulating the deliberate release of GEOs in the environment stimulated debates between ecologists and molecular geneticists. The regulations themselves activated additional disciplinary divisions. I will give two examples of how scientific differences are accommodated to regulatory interests. Debates in basic science can persist for years without any externally imposed requirement that they be settled. Policy debates, however, although they rest on uncertain scientific knowledge, may demand resolution by the political forces of society.

Regulation of a new field of technology has certain characteristic requirements. One of these is that the potential risks be classified in some ascending order. More regulatory focus can then be placed on applications of the technology with "higher" potential risks. This was the strategy for addressing the risks of potential laboratory hazards with rDNA experiments in the mid-1970s.

In its effort to build a taxonomy of risk for GEOs, the regulatory community created the concept of "new organism." Microbes that qualified under the definition would be given special review. The criteria for risk assessment and the jurisdiction of the relevant federal agencies were described in the "Coordinated Framework for Biotechnology Regulation" published in the *Federal Register* on June 26, 1986.[50]

The report distinguished between closely and distantly related organisms. Behind the distinction is the assumption that the higher-risk experiments are those involving genetic exchanges across large evolutionary distances.[51] The reason given in support of this view is that two organisms from the same species or genus have had a greater opportunity to exchange genetic information under natural conditions. Therefore, it is postulated, what takes place in the laboratory is probably a duplication of natural evolutionary exchanges. The same cannot be said of distantly related organisms.

A similar argument was advanced in 1975 during the Asilomar meeting at which geneticists constructed a risk assessment framework for rDNA research that guided a decade of regulation.[52]

Following the putative relationship between risk and evolutionary distance, the EPA classified genetic exchanges as either intra- or intergeneric. The former means that genes are taken from species of the same genus, while the latter refers to exchanges of species of a different genus. Many molecular geneticists supported a risk taxonomy based upon this distinction so long as their own research was not stymied.[53]

There is an interesting parallelism between ecology and genetics on this issue. At the macro level ecologists posit greater risks from introducing organisms into new environments or introducing GEOs into the environment. Whenever the biological systems have not had a chance to co-evolve, the assumption both at the micro and macro level is that discontinuities are more probable.

Bernard Davis provided a different perspective from which to view the question of risk and evolutionary distance. He contends that a Darwinian perspective predicts that mixing genes of distantly related organisms is safer than mixing those of closely related ones. He reasons as follows: The genome is an integrated unity whose elements are co-adaptive. It is like a finely tuned mechanism with highly specialized parts. When a foreign gene is introduced, the likelihood is that it will not integrate into the system. It will slow it down, create an incoherence, or simply clog up the works. Davis concludes that the more distantly related are the organisms in a genetic exchange, the less viable the product will be in the environment.

In his view, the Darwinian perspective implies that there are limits on how far scientists can go in remaking biological species. We can use genetic engineering "to broaden the range of modified organisms for domestication, but not to create radically new or spreading life forms."[54] This analysis was supported by a general principle cited by Hartl that the insertion of any alien genetic material into an organism will disrupt the integration of the genome and reduce its ability to compete in the environment.[55] The greater the evolutionary distance, the more alien the genes. One response to Davis is that his interpretation of genetic exchanges may be correct most of the time, but in some fraction of these events where the organisms are distantly related the effects could be catastrophic.

Criticism of the regulatory distinction between inter- and intrageneric genetic exchanges came from several quarters. Microbiologists noted that the grouping of bacterial species is somewhat arbitrary in the first place. And plant geneticists cited the creation of weeds from nonweeds by intrageneric hybridization.[56]

A second example illustrating the use of a "scientific construction" for an expedited review refers to the type of DNA sequences that are added to the organism. Some DNA sequences serve as the templates for protein synthesis. These are called "coding sequences." Alternatively, other DNA sequences that perform functions except coding for protein synthesis are called "noncoding sequences." The EPA and other agencies excluded from the definition of "new organism'' foreign genetic transfers involving noncoding sequences. This simplified the review process on the presumption that the transfer of noncoding sequences would pose no increased risk to human health or the environment beyond that of the host organism.

Once again, criticism mounted. Enhancement of protein production is an important goal of many commercial applications of genetics and noncoding sequences may regulate the production of a protein. Some ecologists argued that by enhancing or reducing gene expression, ecologically important relationships between an organism and its environment may change. For example, microbes may utilize significantly greater amounts of a substrate and change the soil ecology. Sharples argued that ''the absence of a protein or the amplification of its production could have profound ecological effects.''[57]

Writing in the *Bulletin of the Ecological Society of America*, Robert Colwell et al. commented on the implications of the noncoding exemption for genetically engineering higher organisms.

Ironically, for higher organisms, an important and increasingly dominant theory of the origin of evolutionary novelties and higher taxa is that mutations in regulatory regions are more important than changes in regions that code for gene products for these major evolutionary shifts. How are we to reconcile the evidence for this theory with the idea that regulatory changes cause shifts in phenotype so minor that no change in risk can occur?[58]

Notwithstanding the unresolved nature of these inter- and intradisciplinary conflicts over the questions of how risk is affected by evolutionary distance, genetic baggage, additions versus deletions of genes, and noncoding DNA sequences, the regulatory review process proceeded.

PARADIGMS AND RISK

Since 1962, when Thomas Kuhn's *Structure of Scientific Revolutions* was first published, a vast literature has appeared in the history and philosophy of science centering on the notion of ''paradigm.'' Kuhn cited divisions between the natural and social sciences that turned his attention to the central queries of his work.

...I was struck by the number and extent of the overt disagreements between social scientists about the nature of legitimate scientific problems and methods. Both history and acquaintance made me doubt that practitioners of the natural sciences possess firmer or more permanent answers to such questions than their colleagues in social science. Yet, somehow, the practice of astronomy, physics, chemistry, or biology normally fails to evoke the controversies over fundamentals that today often seem endemic among, say, psychologists or sociologists.[59]

Kuhn appropriated the term ''paradigm'' to describe the universally shared principles of a field that ''for a time provide model problems and solutions to a community of practitioners.''[60]

Much of the scholarship following this work focused on the interpretation of scientific change. Meanwhile, the term "paradigm" emerged as a concept of extraordinary richness in calling attention to the problems of communication across theoretical boundaries. Several key features of the Kuhnian analysis are:

• The most important rivalries in science can be traced to paradigm conflicts.

• Competition between paradigms cannot be resolved by an appeal to the scientific method. "The transfer of allegiance from paradigm to paradigm is a conversion experience that cannot be forced."[61]

• Interparadigm discourse is obfuscated by alternative meanings of core concepts and a disagreement about a priori assumptions.

• New factual information does not resolve the divisions between paradigms since a paradigm is a lens through which one sees the world, including the "facts" of experience.

The history of science provides a vast reservoir of examples illustrating these ideas. They include scientists, separated by centuries, studying the same phenomenon, and contemporaries in the same discipline interpreting experience through alternative conceptual schema. There are also examples of scientists operating within a single discipline resorting to two distinct theories in response to an anomaly in one. For example, Niels Bohr responded to Einstein's classic "photon box" thought experiment that showed inconsistencies in quantum theory by appropriating equations from the General Theory of Relativity. This effort on Bohr's part to save quantum mechanics from paradox by mixing scientific paradigms was met with criticism.[62]

In the case before us, a socially defined problem (technological risk) is brought to the attention of science. Two centers of disciplinary activity address the problem and this brings into sharp relief their different modes of analysis and applications of biological theories. It is instructive to apply the concept of "paradigm conflicts" to the debate between ecologists and molecular geneticists over GEOs. A set of queries are immediately suggested. What distinctive features of theory and practice distinguish the paradigms of molecular genetics and ecology and what is their bearing on the interpretation of risks? What body of theory is shared but applied or interpreted differently to the problem of deliberate release? How are facts viewed differently by each paradigm? What factors impede interparadigm communication?

I shall start by examining some characteristics of the two scientific disciplines. Molecular genetics, which began with the phage research programs and the application of X-ray crystallography to the study of protein and DNA structure, has a strong link to physics. Three Nobel Prizes in biology and medicine were awarded to individuals trained as physicists (Max Delbruck, Francis Crick, and Walter Gilbert) who turned their skills to biological problems.

The lineage between physics and molecular genetics may explain similarities in worldview and philosophy of nature that derive from theories in the two disciplines. Physics is, after all, predictive and deterministic (strongly at the macro level, and statistically at the micro level). As a science, it has come to symbolize the quintessence of the Baconian doctrine of power emanating from knowledge. In physics one can study nature by controlled laboratory experiments. There is an enduring faith that the natural laws may be expressed, as Galileo had proclaimed, in the book of mathematics. Because of its technological accomplishments and quantitative elegance, physics has been awarded the status of high science. Those disciplines, like molecular genetics, that build on its accomplishments and methodology also share in its ethos and hubris.

How different are the roots of ecology. It derives from the empirical traditions of the eighteenth- and nineteenth-century naturalists, animal behaviorists, and evolutionary biologists. Explanations rather than predictions predominate. The emphasis is on systems (communities, ecosystems, nutrient cycles) rather than elements, interactions rather than reactions, and change rather than stability. Even as certain sectors of ecology are becoming more quantitative, laboratory-based rather than field-oriented, and model-fixated, there is a healthy skepticism and pervasive modesty regarding the power of ecological models to predict rather than explain; there is also an aversion to any mechanistic view of causality in biological systems. Herein lies the rudimentary basis for alternative philosophies of nature between ecology and molecular genetics that shape their distinct paradigms. They may have available the same facts, but imbed them in a different notion of causality and explanation. The forms of genetic explanation and ecological explanation usually do not clash since they are applied to different domains of experience. The issue of technological risk, requiring contributions from both disciplines to a single domain of experience, has been a source of conflicting outcomes. The study of paradigm superposition is not well represented in the history or sociology of science. These conflicts within science that I call "scientific disorders of the second kind" open up new avenues for exploring the richness of the Kuhnian analysis.

Part III

Social Controls

9

Human Genetic Engineering: New Ethical Frontiers

> It is now possible by modern technology to make chimeras in which new genes are inserted and effectively expressed in animals. My own belief is that this technology will be used effectively to treat a number of inherited diseases of man.
>
> Martin Cline, 1981[1]

> Drugs may be developed by researchers to correct deficiencies within genes that cause genetic errors, and perhaps defective genes may someday be replaced by healthy ones.
>
> James D. Watson, 1990[2]

The public discourse over biotechnology may be visualized as radiating from one seminal event—the recombinant DNA controversy of the mid-1970s. This singularly influential episode gave form to a constellation of issues that were subsequently cast as products and processes of the new genetics, including: the manufacture of biological weapons, genetic screening and discrimination, the creation of transgenic animals, genetic modification of human ova, microbial effluent from industrial applications, the deliberate release of genetically engineered organisms into the environment, and human genetic engineering. Concerning each of these areas of public concern one may ask:

- Are there important new public problems arising from innovations in biotechnology that call out for regulatory action or some other form of collective response?
- Can our existing laws, regulations, and institutions address the varied issues raised in the public arena?
- What institutions or authorities are responsible for the regulation or oversight of these issues?
- What role does the public play in contributing to the decisionmaking process?

Among the issues previously outlined, some have matured to the point where considerable government presence has already begun. Other issues are still in their embryonic stage of public awareness and debate.

Some problems will never reach beyond a brief burst of media attention; others will activate litigious debate and aggressive legislation. For example, the international Biological Weapons Convention prohibits its signatories from developing organisms for offensive purposes.[3] On May 22, 1990, President George Bush signed into law the Biological Weapons Anti-Terrorism Act of 1990, which makes it a civil crime to violate terms of the Biological Weapons Convention.[4] Environmental applications of GEOs have drawn considerable public debate and congressional study, but no new legislation. The policy issues associated with transgenic animals, cast mainly in legal and ethical terms, have been decided primarily by the courts with only modest congressional initiatives. One area of applied genetics where social oversight has probably not kept up with the intensity of public reaction is human genetic modification. The Department of Health and Human Services (HHS) regulates human experiments involving the modification of genes. However, HHS's role is constrained by two important factors. First, the agency's authority is limited to institutions that receive federal funding. And second, the values of HHS are very much in consonance with high-technology medical interventions.

The chapters in this third section of the book focus on the needs and prospects for regulating different uses of genetic technologies. This chapter covers the genetic modification of humans, human embryos, or human ova; Chapter 10 examines initiatives for developing an integrated and coordinated framework for federal regulation of biotechnologies, with special attention given to environmental applications. Finally, Chapter 11 looks at the role of technology assessment in an effort to develop an anticipatory response to developments in biotechnology.

CONCEPTUAL CLARIFICATIONS

Few issues in the field of biotechnology produce as strong a public reaction as the prospect of human genetic engineering. For some individuals, the mere mention of the term conjures up images of a genetically designed race of humans as depicted in Aldous Huxley's 1932 novel *Brave New World*. The idea that some group of humans would set themselves up as the architects of our genetic makeup plays out as some kind of demonic plan whose adherents, should they exist, must be morally deranged.

But the term "human genetic engineering" (HGE) has a broader range of meanings than what is signified above as a centralized program of eugenics applied to fertilized human eggs. The most generalized meaning of HGE is any controlled intervention designed to affect the constitution or function of one or more human genes either directly in the mature organism or indirectly in human germ cells. This definition includes a number of activities that are summarized as follows.

1. Use of chemicals to alter gene function in a human, human fetus, or fertilized egg.
2. Transplantation of genes in the somatic cells of humans (all cells other than germline or reproductive cells).
3. Cloning of a human being.
4. Transplantation of genes to the germ cells of humans (sperm or egg cells; *in vivo* or *in vitro*).
5. Deletion or addition of genes to the fertilized human egg prior to implantation or gestation.
6. Deletion or addition of genes to the human fetus.

The ethical questions surrounding HGE are contingent on which of its several meanings is being advanced. Human gene therapy applies to those uses of HGE that are aimed at curing disease. Proposals introduced by members of the scientific community generally substitute the term "gene therapy" for "genetic engineering" in order to cleanse the term of any negative connotations.[5] Notwithstanding such efforts at conceptual "sterilization" however, the two terms are still used interchangeably by the media and in popular discourse.

Somatic cell therapy (2) is distinguished from germline therapy (4) in that the latter change, unlike the former, can be passed on to future generations. Eugenics is a concept that is much broader than HGE. It means "cleaning up," "purifying," or improving the human gene pool.[6] Genetic engineering potentially can be used to carry out a eugenics program. But there are many other paths to eugenics, including a national policy of selective breeding as well as the widespread use of genetic screening in conjunction with selective abortion.

Sometimes eugenics is defined as "the science of breeding," while elsewhere it is understood as a national policy directed at "purifying" the race. Individual decisions unrelated to a pattern of state-controlled procreation for the purpose of "improving the species" generally do not qualify as eugenic activities. However, if the state supports or mandates genetic screening (amniocentesis) for identification of certain traits of the fetus (mongolism, Tay Sachs), and encourages or provides incentives for abortion when one or more traits are discovered, then a eugenics policy is being advanced.

Certain forms of human genetic engineering have nothing to do with eugenics whether or not these activities are supported by the state. Gene therapy of human somatic cells that does not affect the germline and therefore will not be passed on to subsequent generations, does not qualify under the term eugenics.

EARLY DISCUSSIONS OF HUMAN GENE THERAPY

The prospect of altering human genes to treat disease was widely discussed by biologists around the mid- to late 1960s. At that time geneticists discovered the mechanisms through which viruses carry DNA to the cells they infect. These discoveries prompted a series of articles on the application of this knowledge to human genetics. In 1966, following a series of symposia and books on the control of human evolution, geneticist Joshua Lederberg, then at Stanford University, postulated for further discussion several ways that genetic engineering might be used to alter our genes.[7] In the area of controlled human evolution, Lederberg distinguished between eugenics, which he defined as selective breeding, and algeny, a term he coined and derived from the fusion of "genetic" and "alchemy," under which he included clonal reproduction (cloning) and genetic engineering.

Robert Sinsheimer is another distinguished biologist who wrote positively about the "designed genetic change" of humankind.[8] He postulated a scenario in which a virus would carry an expressible gene that could cure a serious form of diabetes. At the time Sinsheimer was optimistic about science taking control of the human gene pool. He saw it as an opportunity to improve upon our flawed species nature.

The larger and deeper challenges, those concerned with the defined genetic improvement of man, perhaps fortunately are not yet in our grasp; but they are etched clear upon the horizon. The old dreams of cultural perfection of man were always sharply constrained by his inherent, inherited imperfections and limitations. Man is all too clearly an imperfect, flawed creature. . . . And to foster his better traits and to curb his worse by cultural means alone has always been, while clearly not impossible, in many instances most difficult. . . . We now glimpse another route—the chance to ease the internal strains and heal the internal flaws directly—to carry on and consciously perfect, far beyond our present vision, this remarkable product of two billion years of evolution.[9]

Ironically, Sinsheimer emerges a few years later as a moral conscience for the biological community on the issue of genetic engineering. His optimism for science expressed so poignantly in his 1969 essay turns to a critical skepticism after the discovery of recombinant DNA technology.[10]

Other scientists cast the promise of altering human genes into political terms. They wrote for a much wider audience. Several biologists felt a sense of responsibility to forewarn society about aspects of their work that presented ethical problems. It was reinforced by the moral condemnation of elite science that took place during the Vietnam War period and the anti-positivistic movements that flourished in the 1960s.

Salvador Luria published his essay "Modern Biology: A Terrifying Power" in the *Nation*—an important venue for New Left analysis and critique—and warned us about the technological imperative: "Once the scientific principles are established, technological application is almost certain to come. Thus faced with the prospect of a new technology, we must ask ourselves as soon as possible whether and how it will be used and what we can do about it."[11]

Despite his Huxlian scenario of how biology might be misused ("We may even be tempted to manufacture many identical copies of a supposedly 'superior' individual''), Luria opposed any form of moratorium to prevent potential misuses of biology. He called for ''rational machinery'' to deal with research and its uses and a heightened moral responsibility among scientists to ''tell society, in a forceful and persistent manner, what science is discovering and what the technological consequences are likely to be.''[12]

The period of the late 1960s and early 1970s was accompanied by an explosion of moral self-examination on the part of biologists. In 1969 Jonathan Beckwith of the Harvard Medical School led a team of biologists in isolating the first gene. At a press conference announcing the feat, Beckwith and his colleague James Shapiro spoke of the frightening possibilities that their discovery raised.[13] Subsequently, Shapiro left science because of his concern that genetic research might be used for immoral purposes.[14] But his departure was not permanent. After a stay in Cuba, Shapiro took a post at the University of Chicago where he continued working in bacterial genetics.

In 1972, reports began appearing in the scientific literature describing the steps required for the modification of a mammalian cell. The prospect of human gene therapy was advanced by successes in the use of viruses as vectors for gene transfer. The more formidable problems included putting the transported gene in the correct site and initiating its expression. Along with the excitement about progress in this field, there were also expressions of ambivalence or caution about the eventual application of human genetic modification.

Theodore Friedmann and Richard Roblin were early pioneers in examining the difficult problem of proposing norms for the use of human gene therapy. Their 1972 *Science* essay with the interrogative title ''Gene Therapy for Human Genetic Disease?'' is emblematic of the early sensitivity that some scientists felt toward this issue.[15] The authors described the first documented human gene therapy experiments that were undertaken in 1958 without publicity or public controversy. Two children suffering from a hereditary enzyme deficiency were treated with injections of a virus. At the time it was believed that the virus carried an essential piece of DNA that, when transplanted into human cells, could put into proper balance the body's enzyme chemistry. Performed without public oversight and prior to the establishment of institutional human subjects committees, the experiment failed to achieve its desired outcome. The authors reported that the first effort at human gene therapy was premature and utilized a technique that was not sufficiently demonstrated on animal models.

In 1971, under stricter controls and more accountability for human experiments, scientists working at NIH infected defective human cells with a virus they hoped would carry a missing gene, the absence of which is responsible for the hereditary disease galactosemia. Without the gene a person is unable to utilize galactose, a simple sugar common to dairy products. A baby with this defect that is not put on a dairy-free diet will become mentally retarded and eventually die. The experiment transferred into human cells a normal bacterial gene that was thought to provide the same sugar-utilizing enzyme as the missing human gene. June Goodfield summarized the results of these efforts in her book *Playing God*.

Now this experiment did *not* supply a normal *human* gene to an abnormal human cell. It supplied a normal bacterial gene that could perform the missing function to an abnormal human cell . . . science had indeed taken a first step toward "repairing the human cell."[16]

John Fletcher discussed one of the earliest collaborative efforts at modifying the genes of an adult individual utilizing viral vectors. Between 1970 and 1973 an American scientist assisted a German physician in treating three German sisters who suffered from hyperargininemia—a hereditary disorder involving a deficiency in the enzyme arginase. Without treatment, individuals build up excessive amounts of the amino acid arginine in the blood, resulting in severe mental retardation. The sisters were infected with the Shope papilloma virus, which the physicians believed would stimulate the production of the enzyme and reduce the concentration of the amino acid. The treatment was unsuccessful.[17]

In their 1972 paper, Friedmann and Roblin frame an ethical discussion around the issue of misusing a promising technology. They defined misuse as the premature application of gene therapy or the application of it for anything that does not provide primary benefit to the patient. The ethical issues are cast in a form that renders them best handled by the medical community, those institutions with closely shared goals, or patient advocates.

Another group of scientists who responded to the prospect of human genetic engineering in this early period focused on broad social and philosophical issues. The essays avoided tackling the specific implications of any new technological development on the grounds that the sensitive experiments were still too far off to address with any degree of specificity. In 1970, after a frog had been successfully cloned, Joshua Lederberg's comments captured this perspective.

Such experiments, with laboratory animals, will surely be very fruitful of basic scientific knowledge if the techniques can be developed. It would also have enormous value in livestock breeding, just as cloning (propagation by cuttings) is a mainstay of horticulture. Until such experiments have been pursued in some depth, with other animals, *it is merely a speculative game to discuss applying such reproductive novelties to man.*[18] [emphasis added]

A rather different approach was taken by MIT biologist Ethan Signer, who viewed scientists as the instrument of the power elite. Without a fundamental restructuring of society's values and institutions, science, including genetic engineering, Signer argued, will be exploited by the powerful and used against the powerless. He wrote about scientific elitism, class interests, and technological hegemony. He contended that scientific objectivity was a myth perpetrated by one class interest on another. "Science can be used in conflicting ways in the interest of different sectors of society, and that scientists, rather than participating only in a disinterested search for truth, are by the research they choose to do implicitly advocating the interests of one sector or another.''[19]

Signer's analysis hit a sensitive nerve among critical mainstreamers—those whose framework of analysis did not include a full dressing-down of the capitalist state. But there were some areas of common ground. Signer argued that scientists bear a responsibility to analyze and communicate to the public how their work will be used. This was one proposal embraced by the bioethics community. He fulfilled his own recommendation that university science courses should consider the social effects of the subject matter. Like other scientists reaching out to society during this period, Signer was more adept at exploring the possible applications of genetic engineering than he was at postulating norms for its controlled use. Without concrete cases of human genetic manipulation for consideration, the realization of public policy regarding its misapplication was reduced to a speculative game. The discovery of recombinant DNA in 1973 brought the possibilities closer at hand and gave rise to a new chapter in the public debate over the ethics of human genetic engineering.

GENE-SPLICING AND HUMAN GENE TRANSFER

As previously noted, the first wave of scientific concern about human genetic engineering came in the aftermath of discoveries that viruses could be used to transfer foreign DNA into mammalian cells. The development of recombinant DNA molecule technology greatly simplified the process of gene exchange by using extra-chromosomal rings of DNA called plasmids as the transporting entities. Once plasmid-mediated DNA exchange was refined, the simplicity of its method made it available to a broad sector of scientists, and not just to virologists. The process promised to be more controllable and predictive than viral-mediated DNA transfer.

The international Asilomar conference was held January 24–27, 1975, to address some of the potential risks of rDNA molecular technology. Organizers of the meeting restricted the agenda to laboratory biohazards since it was those concerns that prompted the call for a voluntary moratorium for certain classes of experiments.[20] From Asilomar came an explosion of media attention. A number of post-Asilomar books by science writers with titles like *Biohazard*, *The Ultimate Experiment*, and *Playing God*[21] gave the public its first glimpse of an arcane controversy about biological hazards and new possibilities for modifying the human species.

Among the first generation of popular genetic engineering accounts targeted for a mass market, the one with the broadest public distribution was *Who Should Play God?* by Ted Howard and Jeremy Rifkin. One of the central themes in this work is that lurking behind genetics research is the ideology of eugenics which, like an opportunistic pathogen, waits for the proper moment to infect society: ''Bioengineering a better, or even 'perfect' human specimen has long been the goal of geneticists. In this regard, genetics, unlike the other branches of science, has always had an ideological wing, which has been as indispensable to its development as the scientific discoveries themselves."[22]

According to the authors, whatever the original intentions and applications of genetic engineering, whatever its scale of use or its intended humanitarian directions, it will be led inexorably by this ideology to creating new forms of human life. Howard and Rifkin make the bold assumption that to take the first step is to permit the final consequence.

The step-by-step process leading to the full-scale engineering of human development can already be anticipated. As an initial step, gene surgeons will use recombinant DNA to introduce new genetic material into a human to correct some simple deficiency or disorder—diabetes, for example. . . . Eventually, our new knowledge of DNA will be used to redirect more complex traits. . . . Finally, true genetic engineering will develop. Certain social groups may receive heightened intelligence or superior health courtesy of the bioengineers.[23]

This argument presupposes that science and the society within which it functions are part of a deterministic process. Any theory that views social behavior through a deterministic lens must be approached with skepticism. But we should not dismiss these ideas entirely without examining several forms of the deterministic thesis.

We may distinguish historical determinism from an absolute fatalism. In the former, the social decision matrix consists of critical nodal points where "free" agents choose from among alternative pathways to the future. As an example, under the historical determinist thesis, once we unleashed the first atomic bomb, we were driven by the iron laws of necessity to a global arms race. Others who are more pessimistic see the final outcome as nuclear war. Had another choice been made at the critical juncture, we might have been spared the massive buildup of nuclear weapons.

This view of historical determinism assumes that societies can select the initial state or starting point (node) along a specific historical path. The node represents the initiating event of some more or less deterministic path leading to a final state. Let us assume for a moment that such deterministic pathways existed in technological developments. In the case of genetic engineering, where would the critical node be? Is it the point at which the first microbe was genetically modified? Might it be the first manipulation of a human gene? Alternatively, is the starting point the discovery of the genetic code in 1953, restriction enzymes in the late 1960s, or recombinant DNA in 1973? Furthermore, if Pandora's box of human eugenics has been opened and the action is irreversible, what is the point of contravention? If it must happen because the critical node has been trespassed, we are left only with the option of informing without the possibility of altering the inevitable outcome.

Historical determinism as described above is not a useful framework for anticipating the direction of science and technology. It has no predictive efficacy. A technology has many potential uses but not all are realized. All too often we are made aware of the abuses, but there are clearly many undesirable things that can be done in the name of technology that are proscribed. On the other hand, it *is* far more difficult to control atomic energy once the atom has been split or to ban biological weapons once they have been created.

The Howard-Rifkin position could also be interpreted as a variant of the slippery-slope argument where the acceptance of some forms of genetic engineering will allow more problematic forms to gain acceptance. Slowly, by small incremental steps, human eugenics may be adopted as public policy. The slippery-slope argument may be framed in a deterministic or probabilistic form. However, it is more convincing as a probabilistic thesis.

Suppose there are n types of human genetic engineering (HGE) scaled such that the first level (HGE-1) is least objectionable and the nth level (HGE-n) is most objectionable (according to current norms), with increasing states of "objectionableness" between 1 and n.

The fatalist position is that we will reach the nth state regardless of what interventions we make once the initiating discovery has taken place. The fatal flaw is in the discovery itself. Once this takes place, human weakness draws civilization to disaster. A deterministic but nonfatalistic position holds that once HGE-1 is permitted, we will inevitably draw closer to HGE-n. In this case, HGE-1 is a nodal point that links a precipitating event to a series of escalating applications of HGE. We can avoid HGE-n only if we avoid HGE-1.

The probabilistic thesis maintains that any decision, say HGE-k (where $k<n$), makes HGE-n more likely, but does not causally determine it. It allows for the possibility of human intervention at any point in the chain. Thus, while there is no law of iron necessity that links the treatment of a thalassemic patient by transplanting genetically engineered cells and the implantation of genes in a human egg, these two events are united by a similar technological process and impelled by similar economic forces and professional motivation. The former event gives shape to the latter event without "determining" it. The probabilistic thesis implies that we are treading on ethically sensitive ground. For this reason there must be clarity about the justification for HGE-k and the ethical boundaries between each class in the series of possible human genetic engineering events.

Strong advocates of the first HGE experiments dismiss the deterministic thesis and give little if any credence to the probabilistic thesis. Each event is treated on its own terms. Some opponents of these experiments reject this form of historical reductionism but put in its place a naive technological determinism. No one has successfully argued that the decision to execute the first group of HGE experiments will have no effect on the next generation of experimental interventions into the human genome. It is not unreasonable to hold the probabilistic view, particularly for those who have no confidence in the moral gatekeepers. The next section describes a highly publicized human gene therapy experiment that shows how fragile the system of oversight is capable of becoming.

THE MARTIN CLINE EXPERIMENT

In July 1980, Martin Cline, a scientist at the University of California at Los Angeles (UCLA), transplanted genetically engineered cells into the bone marrow of two women.[24] One was a 16 year-old Italian and the other a 21 year-old Israeli. Both were stricken with an inherited blood disease called beta thalassemia major. The experiments were performed in their respective countries at the Haddasah Hospital in Jerusalem and the University Poly Clinic in Naples. Each woman had some bone marrow cells extracted and treated with a corrected gene for hemoglobin and a gene designed to augment the cells' production when they were transplanted back into the patient.

These experiments represent the first use of recombinant DNA techniques on human beings. The story was followed by the media for a period of months during that year. The ethical issues surrounding the case were complicated by the following factors.

- UCLA refused Cline permission to do similar experiments in California on the grounds that more animal research was needed.
- Permission had been granted by hospital officials in the two countries.
- Patients gave their informed consent.
- The two women had a life-threatening form of the disease that brings death to most patients by their early twenties.
- Cline's work was supported by the NIH, which had guidelines on the use of rDNA techniques. The guidelines were applicable to projects done outside the United States if they were funded at least in part by the NIH.

When the story broke there was a rash of questions. Was Cline morally justified in carrying out this experiment? Did he violate federal regulations? Were patients and hospital officials adequately informed about the nature of the experiment? Should U.S. guidelines for experimental therapies apply to other countries? The moral issues were obfuscated in a cloud of negative publicity. The general criticism of the two experiments fell into four categories.

1. The experiments were premature. There was not enough evidence to conclude that the experiment had any chance of working. The animal data were not deemed satisfactory. The experiments were not designed to provide useful information if they failed. And too little was known about the side effects. Some speculated that placing genetic material in a cell could cause cancer or create new disorders, making patients more ill than they were before the treatment.

2. The experimenter violated the NIH rules, albeit on a narrow technicality.

Cline had approval from federal authorities to insert two genes separately into human cells. However, he went a step further and introduced them in combined form (a recombinant molecule). Cline argued that the separated genes tend to recombine within the cell naturally so that there is no substantive difference between the approved experiment and the one he actually conducted. In response to an NIH investigation Cline described his decision:

I decided to use the recombinant genes because I believed that they would increase the possibility of introducing beta-globin genes that would be functionally effective, and would impose no additional risk to the patient. . . . I made this decision on medical grounds.[25]

Cline decided on his own to alter the protocols of the Israeli experiment and inject rDNA-altered cells into humans in violation of the January 1980 NIH guidelines. The collaborating physician at Haddasah Hospital stated that Cline used recombinant DNA "without letting anyone here know this before, or even whilst the experiment was carried out."[26]

3. The experiments violated the protocols on human subjects. Strictly speaking, the guidelines for the protection of human subjects, initially adopted in 1974, and modified thereafter, apply to investigators engaged in federally funded research at U.S. institutions. But investigators at a U.S. university engaging in human experiments at a foreign institution must obtain approval from the Institutional Review Board or its equivalent at both their host and collaborating institution. Thus, it was more than a moral obligation of the investigators to follow the U.S. guidelines abroad. On this point Cline commented:

I greatly regret my decision to proceed with the use of recombinant molecules without first obtaining permission from the appropriate committees. Although there are extenuating circumstances relating to the patient's clinical condition and to a greater possibility of a successful outcome with recombinant molecules, it is my feeling that I exercised poor judgment in failing to halt the study and seek appropriate approval.[27]

4. The experiment was a step in the direction of onerous forms of human genetic engineering. This final argument was not part of the dialogue that took place in the scientific community, but formed part of the misplaced antieugenic rhetoric.

When the negative publicity over the Cline experiment erupted, there was great concern among members of the biomedical community that a few premature and unworthy experiments might bring a rash of new laws restricting this line of inquiry for the treatment of disease. They felt it was time for scientists to reach a consensus about human genetic engineering and then to educate the public on the results of the consensus. Both a presidential commission report and a National Academy of Sciences study provided the framework for further discussion and public outreach.

SCIENTIFIC CONSTRUCTION OF THE PROBLEM

A meeting was convened at the Cold Spring Harbor Laboratory in 1982 at which 45 biomedical scientists discussed the scientific and ethical issues associated with gene therapy. The proceedings of that meeting, titled *Gene Therapy: Fact and Fiction*, were published as a public information report.[28] The report advanced two principal ideas. First, many human diseases are genetic in origin and cannot be treated by conventional approaches (i.e., drugs, surgery, diet, or environmental factors). Second, if human gene therapy becomes possible it will be an important part of the physician's treatment regime.

The majority of those gathered at Cold Spring Harbor accepted these premises and were optimistic about the new form of medical intervention. The ethical issues were reduced to the queries: How close are we to clinical trials and how do you decide when you know enough to go ahead and act?

When is it justifiable to use conceptually new, untried, and possibly dangerous procedures on human subjects, procedures that are aimed at developing therapy for them when no previous therapy has existed?[29]

Is it ill-advised to use an experimental procedure on a patient when it is unlikely to benefit that patient, but may provide breakthroughs for future cases? Similar questions were asked during the period when experimental organ transplants were initiated. Conventional ethics does not offer an unambiguous response to this query. One tradition in ethics holds that people should be treated as ends and not as means. Using people as experimental subjects solely for the purpose of helping others is an example of the latter and therefore warrants ethical condemnation. But what if these individuals are fully informed and wish to serve as experimental subjects for the benefit of others—a form of philanthropy?

While these issues were raised, the scientists' primary concerns were that research in human genetic engineering not be derailed and that information about human gene therapy must be appropriately managed to avoid public overreaction. The frame of reference, after all, was the public's response to the issue of recombinant DNA research in university laboratories. That convinced many scientists to exercise much greater caution in communicating risks and in bringing to a public forum the dissonances that are expressed within their professional societies.

David Baltimore's response to a query about what is permissible is a striking example of the "we and they" discourse that pervaded the meeting, and the translation of ethics to political expediency.

We have to be extremely careful about what is tried in humans. This is crucial because any attempts at human therapy that are less than totally defensible on scientific grounds are going to muddy the political atmosphere in which we do our science.[30]

Biomedical scientists sought ways to depoliticize human genetic engineering despite its powerful political and social symbolism. Several strategies were advanced to achieve this outcome. First, the concept of human genetic engineering was subdivided into two parts, one bearing a medical association and the other a political one. This was accomplished through the distinction between somatic cell and germline genetic engineering. The former was identified as medical therapy, while the latter was seen by some as a form of eugenics. Advancing this distinction was important for minimizing the impact of the slippery-slope argument and for undercutting the deterministic thesis, both of which were gaining adherents. Among some scientists, there was skepticism about whether people could be taught to accept the distinction.

People will never understand the difference between gene therapy of somatic cells and modification of the germ line. That is a fundamental distinction. We all understand it and we have no difficulty with it. But although we will keep trying—we have to keep trying—I believe we will never actually get it through to the general population because these concepts are too close to each other, and they require too much scientific understanding to make those distinctions.[31]

It is not complex science that fixes this distinction, but rather a concept quite accessible to a lay audience. In somatic cell HGE, the subject alone is modified by the procedure. In germline HGE, the changes are carried through multiple generations. Pollsters used the distinction quite handily in attitudinal surveys on human genetic engineering.[32] (See the last section of this chapter on public opinion.)

A second approach for depoliticizing science is to reconstruct the terminology. Words carry political messages, some positive and some negative. The earlier genetics debates provide a few examples. The term "breaching species barriers" sheds a negative light on recombinant DNA research. It projected a scientific hubris that played poorly on people's sensibility. Eventually, this terminology lost favor among scientists.[33]

To cite another example, the term "mutant" has a common usage in biology, as in "mutant organism" or "mutant genes."[34] During the community debates over the field test of ice minus, scientists railed against citizens and some newspapers for using the provocative term "mutant bug" instead of the more neutral expression "genetically modified organism."[35]

Similarly, when the use of HGE was being considered as a treatment for human disease, the term "therapy" and not "engineering" was deemed appropriate. This usage would reinforce the public's positive associations with biomedical science.

A third effort at depoliticizing HGE was aimed at dramatizing the idea that its therapeutic use was the *only* available cure for a dread disease. Powerful emotional overtones accompanied the message that a patient, holding on to life by a thread, might be offered a reprieve with gene therapy. This imagery overshadowed any political baggage or ethical liability that might be carried by the procedure.

While the biomedical scientists and medical practitioners joined forces to build a solid case for human genetic engineering, the issues were eventually addressed by broader constituencies of bioethicists, theologians, and lawyers. One of the most significant meeting grounds for this confluence of interests came in the form of a presidential commission.

THE BIOETHICISTS' FRAMEWORK

On June 20, 1980, four days after the Supreme Court decision on the patenting of a microorganism, the general secretaries of the National Council of Churches, the Synagogue Council of America, and the United States Catholic Conference wrote President Jimmy Carter warning of the potential misdirections in genetic research. "History has shown us that there will always be those who believe it appropriate to 'correct' our mental and social structures by genetic means, so as to fit their vision of humanity. This becomes more dangerous when the basic tools to do so are finally at hand. Those who would play God will be tempted as never before."[36]

The President's Commission for the Study of Ethical Problems in Medicine and Biomedical and Behavioral Research had already been established by the Carter administration when the letter from the three general secretaries was received. However, it wasn't until the letter was publicized that the commission was requested by the president's science advisor to undertake a study of the impact of rDNA and related technologies on humans. That study began in September 1980 and continued for two years.

A few months prior to the release of the commission's report, in July 1982, Nicholas Wade wrote an editorial for the *New York Times* titled "Whether to Make Perfect Humans." The editorial concluded with the ominous warning: "Deliberate manipulation of the human germline will constitute a watershed in history, perhaps even in evolution. It should not be crossed surreptitiously, or before a full debate has allowed the public to reach an informed understanding of where scientists are leading. The remaking of man is worth a little discussion."[37]

The commission's report titled *Splicing Life* was issued in November 1982. The commission's analysis followed the convention of distinguishing between noninheritable and inheritable forms of HGE. In the former (somatic gene therapy), when the treatment is directed at a disease, the commission concluded that there was no justification for distinguishing it from other forms of medical treatment. Two important provisos were added. First, it was recognized that the use of HGE on somatic cells could be an irreversible therapy. Second, in an effort to modify or replace somatic cells, germ cells might also be altered.

The commission cited with caution and treated as problematic HGE aimed at enhancing "normal" people as opposed to remedying a recognized genetic defect.

Interventions aimed at enhancing "normal people," as opposed to remedying recognized genetic defects, are also problematic, especially since distinguishing "medical treatment" from "nonmedical enhancement" is a very subjective matter; the difficulty of drawing a line suggests the danger of drifting toward attempts to "perfect" human beings once the door of "enhancement" is opened.[38]

However, the commission shied away from any prohibitory language. It simply drew attention to the heightened ethical concerns: "Altering the human gene pool by eliminating 'bad' traits is a form of eugenics, about which there is strong concern." The closest the commission came to expressing unqualified opposition to an HGE application was the prospect of creating human-animal hybrids through crossing germlines. This proved to be the most emotionally charged aspect of the report.

Could genetic engineering be used to develop a group of virtual slaves—partly human, partly lower animal—to do people's bidding? . . . Dispassionate appraisal of the long history of gratuitous destruction and suffering that humanity has visited upon the other inhabitants of the earth indicates that such concerns should not be dismissed as fanciful.[39]

Critics were quick to attack the passage as far-fetched and as providing grist for the zealots who view any form of HGE as part of an inexorable path toward the most extreme uses.

At best, the commission illuminated the gray zones, but provided no operational framework for marking the boundaries between acceptable and unacceptable uses of HGE. On the overall question of whether human genes should be modified, the report could find nothing inherently inappropriate with such applications. In those areas of HGE that may be scandalously abused (e.g., control of sexual reproduction on a large scale), the commission concluded that we must have strong, democratic, and vigilant institutions along with a commitment to individual rights. *Splicing Life* was a landmark study that provided a serious forum for many perspectives. However, it brought us no closer to resolving the delicate issues of enhancement therapy and germline experimentation.

Meanwhile, breakthroughs in clinical uses of gene therapy were beginning to appear in the press. Scarcely one month after the release of *Splicing Life*, in December 1982, the *New York Times* ran a front-page story under the byline of veteran science correspondent Harold M. Schmeck, Jr., with the headline "Activity of Genes Reported Altered in Treating Man," and a subtext that read "Test termed a demonstration that this discipline has come to the bedside." The story cited a new study in the *New England Journal of Medicine* by scientists from the NIH and the University of Illinois College of Medicine which reported a successful attempt to alter gene activity by the use of chemicals. The report

centered on one patient stricken with the inherited blood disorder beta thalassemia, which results in the body's production of abnormal hemoglobin. The
scientists used a drug that reactivates dormant genes by eliminating a key chemical blocker. A total of five patients were reported to have benefited from the
treatment.[40]

The story brought public attention to the frontline clinicians who were seeking
new tools for the relief of life-threatening inherited diseases. Nevertheless, the
concerns over enhancement and alteration of the germline had not diminished.
Jeremy Rifkin, capitalizing on the mood of ambivalence over HGE, sent out a
ten-page epistle on the moral arguments against genetic engineering of the human
germline to religious leaders, scientists, theologians, and bioethicists asking if
they would support a moratorium on such experiments. His letter stated:

It will soon be possible to engineer and produce human beings by the same technological
design principles as we now employ in our industrial processes. . . . In deciding whether
to go ahead or not with human genetic engineering we must all ask ourselves the following
question. Who should we entrust with the authority to design the blueprints for the future
of the human species? . . . Who do we designate to play God? The fact is, no individual,
group or set of institutions can legitimately claim the right or authority to make such
decisions on behalf of the rest of the species alive today for future generations. Genetic
engineering of the human germline cells represents a fundamental threat to the preservation
of the human species as we know it, and it should be opposed with the same courage
and conviction as we now oppose the threat of nuclear extinction.

Rifkin garnered 57 supporters spanning a broad political spectrum, including
notables like Jerry Falwell of the ultraconservative Moral Majority and Avery
Post of the liberal United Church of Christ. These leaders represented a constituency in the tens of millions of people.[41] By soliciting signatures and publicizing a single list of uncommon bedfellows, Rifkin had skillfully orchestrated
a media event. Moreover, Rifkin had taken one step beyond *Splicing Life* in
trying to find a consensus for a clear and distinct forbidden zone. This action
reactivated a debate among biomedical ethicists, theologians, and scientists on
the proper moral and legal boundaries for HGE.

FRACTIONALIZATION OF HGE

Modern science follows rather judiciously Descartes' principle that large problems are solved by dividing them up into manageable parts. The same method
has been applied to ethical problems. Rather than viewing human genetic engineering as a cluster definition, scientists and bioethicists fractionated the concept, explicating its subtle components. At first, HGE was bifurcated into somatic
cell and germline manipulation. When that proved to be inadequate, finer divisions were created. By treating each division of the concept as a special
category, the concerns about a slippery slope can be more readily managed.

The National Academy of Sciences and its affiliate the Institute of Medicine sponsored a workshop on human gene therapy in October 1986. Two years later a book based upon these meetings and written by medical/science writer Eve Nichols was published under the academy's imprimatur.[42] Following conventions discussed at the meeting, the book disaggregates HGE into four categories of experiments.

1. Somatic Cell Gene Therapy
2. Germline Gene Therapy
3. Enhancement Genetic Engineering
4. Eugenic Genetic Engineering

The principal division is between gene therapy and genetic engineering, where the former connotes the curing of disease and the latter the "improvement" of human beings. Somatic cell gene therapy is defined rather narrowly as "the insertion of a single gene into the somatic cells of an individual with a life-threatening genetic disease."[43] This restricted form of HGE is cited by Nichols as "the only technique now considered appropriate for use in human beings."[44] Germline therapy is defined as the insertion of a healthy gene into a human egg with a specific genetic defect. This definition builds on a controversial distinction between healthy and unhealthy genes. Are extremely short-statured people endowed with an "unhealthy" gene? Are people who are hypersensitive to certain toxic chemicals endowed with an "unhealthy" gene? Efforts to rid the species of "unwanted traits" that are not associated with disease are referred to by ethicists as positive eugenics.

The third category employs the term genetic engineering rather than gene therapy. The implication is that the intervention has moved beyond medical treatment. By "enhancement genetic engineering" Nichols means the alteration of a specific characteristic of a human being "in a preferred direction." These characteristics would be altered in the fully formed person through modification of somatic cells.

The techniques for the attainment of (1) and (3), namely somatic cell HGE, may be equivalent. The categories are not distinguished by the science involved but rather by the social/medical objectives. In (1) the objective is the elimination of the clinical manifestations of a disease, whereas in (3) it is the improvement of a clinically normal individual. This distinction is easily blurred. Medical abnormalities come in all forms from mild to severe. Many abnormalities are not medical although they might be interpreted as such. Some phenotypes in conjunction with environmental factors are associated with a higher likelihood of clinical pathologies. And there are cases where the medical community is not in agreement on whether a physical condition should be characterized as a disease (e.g., fibrocystic breasts).

The final category, eugenic genetic engineering, is defined as the alteration of complex traits involving multiple genes, such as intelligence and organ for-

Table 9.1
Nichols–NAS Typology on Human Genetic Engineering

	Experimental Subject	Target Cell	Single or Multiple Genes	Cure or Enhancement
1. Somatic Cell Gene Therapy	persons	somatic	single	cure
2. Germ Line Gene Therapy	eggs/fetus	germline	single	cure
3. Enhancement Gene Engineering	persons esp. children	somatic	single/multiple	enhancement
4. Eugenic Gene Engineering	eggs	germline	multiple	enhancement

mation. This definition of eugenics is far narrower than the more common usage where it signifies "cleaning up" the gene pool of defective or inferior genes (negative eugenics) or, conversely, improving the gene pool by increasing the frequency of superior genes (positive eugenics). It is left unclear whether this form of intervention is carried out on fertilized eggs (including zygotes and fetuses) or persons, although it is most likely the former.

The four-part classification developed by Nichols in cooperation with NAS is unmistakably opportunistic (see Table 9.1). It stipulates new definitions that treat one class of experiments (type 1) as unambiguously permissible. But the only distinction that is defensible on scientific grounds is that between somatic and germline HGE. The boundary between enhancement and cure is socially constructed. Positing well-defined categories of illness and normalcy trivializes the gray areas and neglects the important distinction between potentiality and actuality in the onset of disease.

EUGENICS

Much of the ethical burden on HGE is carried by the prospect of altering human germline cells. Government reports and science-based studies acknowledge the special moral context of such interventions. In many of the sources I have cited, the possibility is left open for future experiments that would alter the human germline. Some advocates of germline engineering defend it on the grounds that a program of negative eugenics is morally acceptable. Simply put: By engineering the fertilized egg, why not rid the human gene pool of those genes responsible for undesirable traits?

The argument against any form of germline intervention rests on two premises. First, there is no natural demarcation between "unhealthy" and "healthy" genes.

There may be broad consensus for cases at the extremes (e.g., the gene responsible for Severe Combined Immune Deficiency or SCID, is deleterious), but
there are other cases where a gene may serve positive and negative roles (the
sickle-cell trait reduces the likelihood of malarial infection).[45]

Second, once a program of negative eugenic engineering gets initiated, research into positive eugenic engineering will be forthcoming. This argument is
based, to some extent, on the first premise. The likelihood of moving from
negative to positive eugenics is increased when the dividing line is blurred. These
and other considerations were taken up during the next critical period, when
oversight for human genetic engineering was assumed by the National Institutes
of Health.

After the publication of *Splicing Life*, the NIH's Recombinant DNA Advisory
Committee set up a study panel in April 1983 to determine whether this committee
should play a role in the review of experiments involving the modification of
genes. A year later the NIH accepted such a role and established a 15-member
Working Group on Human Gene Therapy (hereinafter Working Group).[46] In
January 1985 the NIH published a set of draft guidelines for gene therapy
experiments developed by the Working Group. After public review and revision,
the final document, titled "Points to Consider in the Design and Submission of
Human Somatic Cell Gene Therapy Protocols," was adopted by the RAC in
September 1985. The protocols were subject to broad public and scientific review.
While focusing on somatic cell modifications, the protocols were mindful of
germline consequences. Investigators seeking permission to do such experiments
were responsible for providing information on the likelihood that the proposed
somatic cell gene therapy would inadvertently affect reproductive cells.

"Points to Consider" made explicit reference to the possibilities of eugenic
uses of genetics. "While research in molecular biology could lead to the development of techniques for germ line intervention or for the use of genetic
means to enhance human capabilities rather than to correct defects on patients,
the working group does not believe that these effects will follow immediately
or inevitably from experiments with somatic-cell gene therapy."[47] According to
the document the RAC would not "at present" entertain proposals for germline
alterations, which left the future somewhat uncertain.

Two citizen initiatives advocated an outright ban on germline experiments.
Jeremy Rifkin's role in publicizing signatures of scientists and religious leaders
supporting such a ban was discussed earlier. The Committee (now Council) for
Responsible Genetics (CRG)—a national public interest group located in Boston—issued a policy position in response to the draft protocols. The CRG's
proposal asked the RAC not to review or approve two types of experiments;
first, human genetic therapy that (1) is not aimed solely at the relief of a life-
threatening disease or severely disabling condition, or (2) could alter germline
cells; second, *in vitro* recombinant DNA experiments that alter human germline
cells or early human embryos. CRG's principal justification for the first prohibition was:

The establishment of restricted zones of application of HSCT [Human Somatic Cell Therapy] at the outset will make it possible to support the most urgent and humane applications of HSCT with less fear that such support gives tacit approval to other uses that, propelled by economic and professional pressures, may slip through the review process.[48]

The CRG also requested the NIH to take a position against the use of HSCT for enhancement of human characteristics, arguing that such uses will result in the medicalization of social problems.

The concept of an ideal height, ideal skin color, or ideal weight is a social construction. Genetic technologies designed to correct normal variations in human phenotypes help to legitimize an ideology that places inferior value on certain expressions of human diversity ...NIH needs to enunciate clear restraints on the use of HSCT for the purpose of enhancement as to avoid a slow and incremental drift toward such applications without adequate social assessment.[49]

This was not a fixed and unyielding opposition but one based on the need for a broad public consensus and wider democratic participation.

The CRG also asked the NIH to make an explicit statement that it would not approve any protocols involving the genetic manipulation of human germline cells through the addition or deletion of genes in the sperm, egg, or zygote. The CRG argued that this would be tantamount to experimentation on future generations with no possibility of informed consent and that it would establish the foundation for a program of eugenics.[50]

The RAC rejected the CRG proposal and decided against making any a priori judgments about a class of experiments that should be proscribed. In a memorandum to the RAC, the Working Group on Human Gene Therapy stated that specific proposals are a useful way to discuss public policy. Therefore, the prohibition of any types of proposals on genetic therapy was deemed detrimental to the general policy discussion. While the Working Group accepted the boundary between somatic and germline HGE, it did not wish to close off a case-by-case review for the latter. "With respect to germ-line cells and embryos this subcommittee has not specifically addressed the complex issues involved but expresses its concern that the CRG recommendation would impede research and clinical advance in reproductive biology."[51] LeRoy Walters, bioethicist, RAC member, and chairman of the Working Group, argued in print that successful somatic cell therapy will give justification to germline therapy. Assume that HSCT is used to cure single-gene defects like cystic fibrosis or sickle-cell anemia. The treated patients will grow to adulthood and will be capable of reproducing. As a result, they are homozygous carriers of genetic disease. Each subsequent generation could be treated or, as Walters proposes, some patients might find it more acceptable to prevent the transmission of the disease by germline therapy.[52]

Another reason Walters gives for pursuing germline therapy is that HSCT might not work. "Early intervention that affects all the cells of the future organism, including the germ cells, may be the only means available for treating cells or tissues which are not amenable to genetic repair after birth."[53]

For the most part, scientists have been cautious about germline therapy, but there is no consensus about establishing a Maginot line between somatic and germ cells. Bob Williamson at the University of London sees a role for embryo therapy in cases like cystic fibrosis where he believes it may be desirable to "alter the inheritance ('genotype') of the individual as well as the way his body works ('phenotype')."[54]

In response to the prospect of human genetic engineering, the NIH preferred to establish procedural recommendations exclusively, leaving open all future possibilities. Any attempt to fix boundaries was opposed by the RAC Working Group on the grounds that it would "impede research and clinical advance in human reproductive biology."[55] Although it was an explicit practice of the NIH not to fund germline manipulation, the agency refused to formalize that practice in response to petitions by public groups.[56]

ELUSIVE MORAL BOUNDARIES

The debate over HGE has engendered three types of responses: those who advocate distinct moral boundaries (such as between somatic and germline, or between therapy and enhancement); those who support pragmatic boundaries (such as what is feasible and acceptable versus what is still debated); and those who are opposed to any articulated boundaries (each experiment is taken on its merits regardless of its domain). Figure 9.1 illustrates how a particular classification scheme sets the framework for policy discussions. With finer distinctions set within the main classification of somatic and germline interventions, the morally suspect applications of HGE have been assigned to narrower and narrower categories.

Harvard bacteriologist and former RAC member Bernard Davis has argued that the ethical concerns of HGE should be directed at any future efforts to improve or "blueprint" people according to somebody's plan. For Davis, the only justified ethical line to draw is that between medical and eugenic uses of genetic engineering, but not between somatic and germline cells. Eugenics, in Davis' terminology, is synonymous with embryo enhancement.[57] The removal or suppression of genes in an embryo associated with some clinical abnormality is not, in this view, a eugenic intervention.

W. French Anderson is a physician at the Laboratory of Molecular Hematology at the National Heart, Lung, and Blood Institute, NIH. On January 19, 1989, the director of NIH approved Anderson's clinical protocol to insert a foreign gene into the immune cells of cancer patients. The technique is identical to those planned for use in human gene therapy.

Anderson, a pioneer in human gene therapy, entered the public sphere for several years engaging ethicists, religious leaders, and the general public in a dialogue that guided him toward a position on the ethics of HGE and cleared a

Figure 9.1
A Classification Scheme for Policy Debates Over Human Genetic Engineering

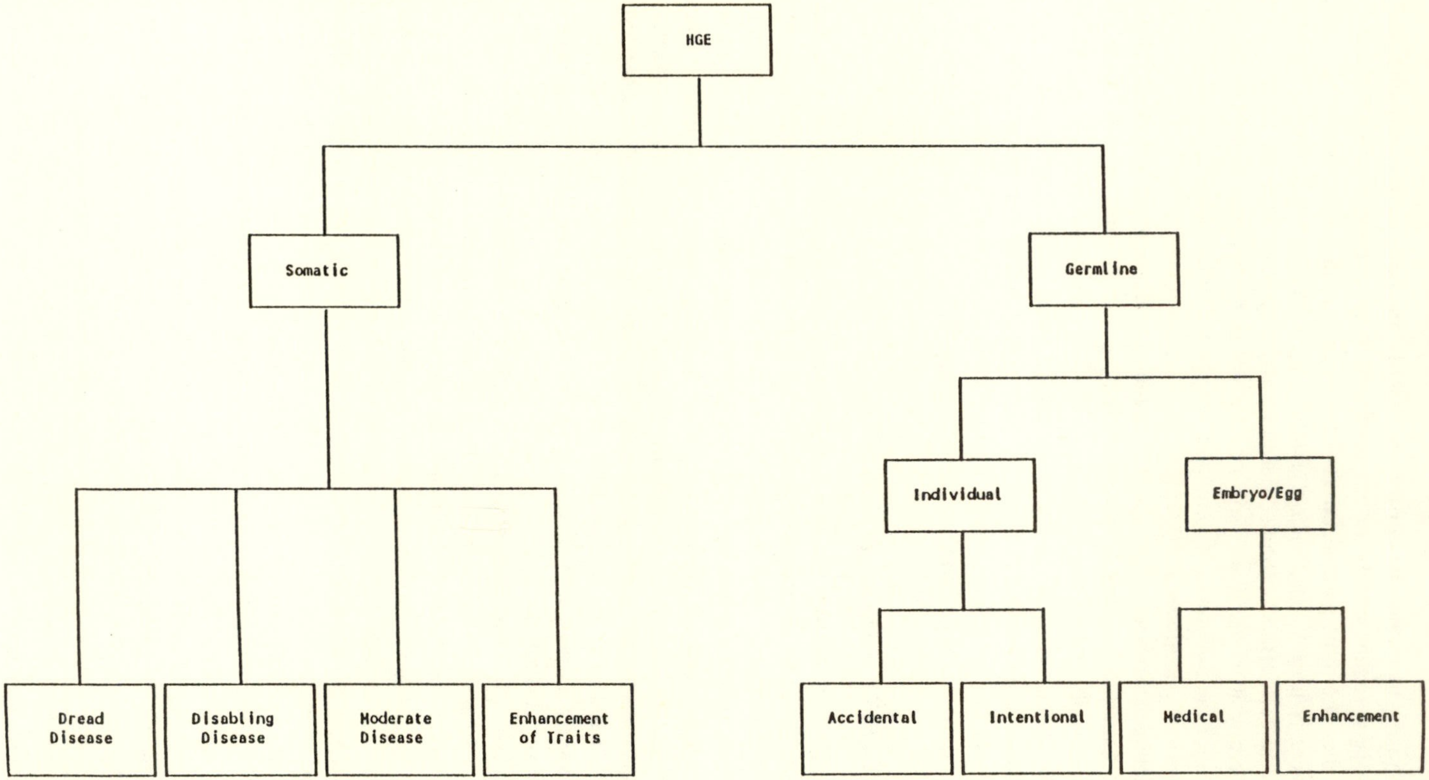

path for the first series of human experiments. He reached a conclusion that a sharp line can and should be drawn to distinguish acceptable from unacceptable forms of human gene therapy. Genetic engineering, he contends, should be used on humans only for the treatment of serious diseases and not for improvements or enhancements of the human condition. Anderson also accepts the use of gene therapy on the germ cells of adult individuals who want their children to receive a normal gene, but only under the conditions that: (1) the effectiveness and safety of the techniques are established first with somatic cells; (2) the techniques are reproducible and reliable; and (3) public awareness and approval is assured.[58]

Anderson has approached the issue of *in vitro* germline gene therapy cautiously in his published essays. While he offers no moral criteria for proscribing human germline manipulations, he does suggest safety and pragmatic grounds. Genetic enhancement therapy, according to Anderson, should be avoided because it would threaten human values by both creating a medical hazard and requiring moral decisions that would lead to inequality and discrimination. Furthermore, he believes that it would be difficult, if not impossible, to establish a boundary once enhancement HGE had begun.

In contrast, David Suzuki and Peter Knudtson define eugenics as "a strategy of trying to orchestrate human evolution through programs aimed at encouraging the transmission of 'desirable' traits and discouraging the transmission of 'undesirable ones.'"[59] They include medical-type traits such as "mutant hemoglobin molecule" in their criteria. The authors advocate a prohibition of germline therapy "without the consent of all members of society." How can we ever obtain the consent of all members of society? Could they possibly mean every last member or simply some majority? There has never been an ethical debate resolved by appealing to the consent of "all members of society." At the very least, their call for the consent of the governed on the moral issues of HGE implores the policymakers to expand the arena of participation. Public attitudes toward HGE will be discussed in the following section.

The positions on HGE of Davis, Anderson, and others are based on a distinction between medical and nonmedical genetic traits. Figure 9.2 provides a two-dimensional matrix of prospective medical and nonmedical applications of HGE. At the extremes are conditions that are sharply defined as medical (bottom) and nonmedical (top). Genetic traits in the middle signify conditions with both social and medical components. For example, phenylketonuria (PKU) is a hereditary disease that is caused by the lack of an enzyme that converts one amino acid (phenylalanine) into another. Without the enzyme and with no treatment, a child becomes brain damaged. Once screened at birth, the disease is successfully treated by a diet low in phenylalanine. However, even in its successful treatment, IQ deficits are observed in older children. Since an IQ deficit in this case is correlated with a medical abnormality (although IQ in itself is not considered a medical problem), PKU falls into a gray zone. From this example, legal scholar Julie Gage concludes:

Figure 9.2
Medical and Nonmedical Applications of Germline Gene Therapy

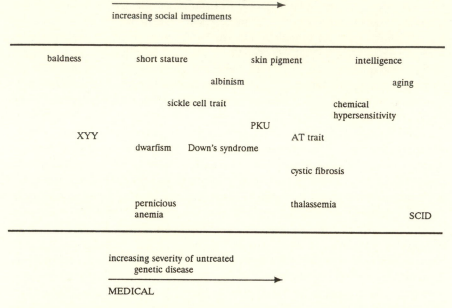

NON-MEDICAL (Enhancement)

increasing social impediments

baldness	short stature	skin pigment	intelligence
	albinism		aging
	sickle cell trait	chemical hypersensitivity	
	PKU		
XYY		AT trait	
	dwarfism Down's syndrome		
		cystic fibrosis	
	pernicious anemia	thalassemia	SCID

increasing severity of untreated
        genetic disease

MEDICAL

The suggestion that gene therapy for enhancement purposes will not follow inevitably from gene therapy to correct defects shallowly skims the issue. There is no easy line between enhancement and curing defects.[60]

Genetic markers can reveal an individual's predisposition to disease. As an illustration, the AT trait (ataxia telangiectasia is a recessive and often fatal genetic disease) predisposes its carriers to a higher risk of cancer. Recognizing that predisposition is a statistical concept, is predisposition on par with having a disease for the purpose of distinguishing between therapy and enhancement?[61] Is preventing a predisposition "therapy" or "enhancement?"

In July 1990 the *New England Journal of Medicine* reported a successful experiment in which elderly men were treated with human growth hormone (HGH). The study indicated that compared to controls, those treated appeared more vigorous and showed physiological signs that belied their age. The body's production of human growth hormone declines precipitously after age 60. One theory maintains that continued production of HGH may slow the aging process. If this can be achieved by turning the dormant growth hormone genes on, would this be considered therapy or enhancement? Is aging a disease?

Without a clear distinction between the medical and social interpretation of a trait or physical state, a consensus position on germline genetic modification does not seem close at hand. This is borne out by public opinion polls.

PUBLIC ATTITUDES ABOUT HUMAN
GENETIC ENGINEERING

Since the early 1970s, scientists have become more attentive to public opinion. The recombinant DNA affair represents a milestone in the establishment of a new dialogue between biomedical science and society. While the issues around human genetic engineering were set aside during the Asilomar conference and the subsequent community debates that took place in cities like Cambridge, Massachusetts, scientists like French Anderson pursuing research in human gene therapy realized that public support was essential to the successful adoption of their program. Some geneticists approached the issue of human gene therapy like a public education campaign, with a number of intact presuppositions. First, they believed the public must be taught to appreciate the difference between eugenics and gene therapy. Second, the public must be helped to understand that each form of human genetic engineering should be evaluated on its own grounds and does not increase the likelihood that some other form, with more sensitive ethical consequences, will be adopted. Third, society can address the ethical issues of concrete possibilities. Unrealizable or remote hypothetical cases should not, however, determine public policy.

Just what is the public attitude toward these issues? Public opinion polls on such matters are quite mercurial; nevertheless, it is instructive to have some baseline. The first major public opinion poll on biotechnology was initiated by the Office of Technology Assessment and conducted by Louis Harris and Associates between October 30 and November 17, 1986. A national probability sample was taken of 1,273 American adults over a 19-day period. The results of the poll were published by OTA as part 2 of a series of special reports titled *New Developments in Biotechnology*.[62]

Over one third of the sample (37%) reported that one or more members of their immediate family had a genetic problem and therefore might perceive a benefit to HGE. Starting with the most general and moving to the specific, pollsters asked about the morality of genetic manipulation. To the general question "On balance, do you feel that changing the genetic makeup of human cells is morally wrong?" 52 percent responded negatively; 42 percent affirmatively; 6 percent uncertain. For those with a college degree the results in favor of HGE were significantly higher: 66 percent said it was not morally wrong while 28 percent said it was; 5 percent were uncertain.[63]

When the pollsters turned to specific cases of gene therapy for the treatment of disease in an individual, the approval response ("strong" or "somewhat") was in the 77–84 percent range. However, when asked about enhancement of traits (i.e., improving the intelligence level that children inherit) the results were dramatically reversed: 53 percent disapproved; 44 percent approved. OTA concluded that "only a minority of the public says it approved the use of human genetic manipulation for eugenic rather than therapeutic purposes."[64]

The poll also asked respondents about somatic/germline applications of HGE. The question was directed at eliminating the genes of fatal diseases.

Suppose someone had a genetic defect that would cause usually fatal diseases in them and would likely be inherited by their children. Do you think that doctors should be allowed to correct only the gene affecting the disease in the patient, only the gene that would carry the disease to future generations, both genes, or neither gene.[65]

The majority (62 percent) voted approval for both somatic and germline intervention. Ironically, more respondents affirmed the right of doctors to eliminate the gene from the gene pool exclusively (14 percent) than curing the patient exclusively (8 percent). A mere 19 percent of those questioned opposed germline therapy for dread disease. This survey informs us that the part of HGE that is particularly troublesome on ethical grounds to the public is enhancement, whether applied to the individual or to the germline.

Based upon this public opinion sample, the critical factor for policy is thus the distinction between using HGE to treat an individual (or eradicate the relevant genes from their gene pool) for a fatal disease, and using HGE for "improving" qualities of clinically normal individuals or selecting qualities for future generations.

But this raises a dilemma. Where distinctions are nebulous, the slippery-slope argument is most applicable. The somatic/germline boundary, while relatively intact, is not the major issue for society. The enhancement/medical boundary is the critical one in the public's mind, but most vulnerable to slippery slope. While gene therapy pioneers like French Anderson posit an ethical window for diseases that produce significant suffering and premature death, there is no philosophical grounding for anything more than a pragmatic boundary that, some argue, will eventually be overridden. Once the right of somatic cell therapy becomes established, it is doubtful that its use can be restricted exclusively to medical treatments any more than surgery can be so restricted or abortion can be prohibited for sex selection.[66]

Currently, the oversight responsibility for human genetic engineering rests jointly with the NIH and local Institutional Review Boards (IRBs). The protocols developed by the NIH apply only to research conducted or sponsored by institutions receiving support for recombinant DNA research from that agency. This method of oversight that combines a modified RAC with the IRB process was favored over other proposals such as a bioethics commission appointed by Congress or a presidential commission.

10

The Growing Complexity of Regulation

Rather than developing positive new initiatives and approaches to the issues raised by this innovative industry [biotechnology], government depends on old and outmoded regulatory structures that have even failed to deal with more traditional environmental and health risk questions.

Congressman James J. Florio, 1985[1]

Unfounded alarm about genetic engineering can lead to unnecessarily stringent regulations, hamstringing U.S. industry. . . . [Biotechnology] is being regulated into submission.

Winston J. Brill, 1988[2]

Most major forms of environmental legislation in the United States have been enacted in the aftermath of a crisis. It need not have been a sudden crisis or catastrophe. The issue may have simmered for years when, for a variety of reasons, it finally reached a crescendo and gained public support for the passage of legislation. No single event is sufficient to account for the change in regulatory climate, but some events evoke such powerful imagery that they symbolize and become identified with the cause.

The Pure Food and Drug Acts were passed after Upton Sinclair published his classic novel *The Jungle*, which dramatized to a popular audience the wretched conditions in the Chicago meat-packing houses. Soil conservation legislation came after the great dust bowl destroyed millions of acres of southwest farmland. International attention to atmospheric nuclear testing was dramatized by the discovery of strontium 90 in mothers' milk. The Clean Air Act was finally passed when national attention focused on the health effects of urban smog. Passage of the Superfund Act was aided by the tragedy of Love Canal, New York, the first internationally recognized case of a community abandoned because of industrial waste contamination. Federal legislation on local emergency preparedness for toxic spills and mandatory toxic emission disclosure by industrial polluters (Superfund Amendments Reauthorization Act, 1986) was catalyzed by the shocking events at Bhopal, India, where over two thousand civilians perished after their exposure to a toxic gas that leaked from a Union Carbide plant.

Many other environmental problems muddle along a path of regulatory incrementalism, as proponents for reform hope for a key event that would mark a major shift in the federal government's environmental agenda. Among such problems are acid rain, global warming, and depletion of the rain forests. Given this tradition in the history of American environmental regulation, where does the regulation of biotechnology fit in?

This chapter reviews the early development of federal environmental regulations for genetically modified organisms, the debates over the appropriate forms of regulation, and federal efforts to resolve problems associated with fragmentation and inconsistency within the regulatory bodies.

GENESIS OF REGULATION

Biotechnology is atypical of other twentieth-century regulatory initiatives. First, the issue grew out of a laboratory setting and was only cast as an environmental issue years later. Occupational exposure of workers to industrial biochemicals, pharmaceuticals, and GEOs was bypassed in the process. In contrast, issues such as the hazards of toxic chemicals gained notoriety first as occupational diseases and second as environmental hazards. Vinyl chloride, lead, arsenic, asbestos, and cotton dust were the symbols of reform for a new generation of occupational health and safety laws. Rarely, if ever, does social regulation start in scientific laboratories and branch out to other sectors (agricultural, industrial, domestic, and occupational). Chemical labs, for example, traditionally have been exempt from federal toxics legislation.

Another feature that distinguishes biotechnology from other regulatory initiatives is that it does not define a characteristic substance, event, or industrial sector. Regulatory concerns are directed at a class of laboratory procedures that may be applied to one or a billion organisms. The techniques may be used to transform any biological species, from the smallest virus to higher primates. Part of the dilemma for regulators is in deciding whether the process or the outcome ought to prescribe the form of regulation.

A third factor that sets biotechnology apart from other areas where environmental regulation has been pursued is that, thus far, no documented hazard has resulted from its use. None of its products has been implicated in human disease or ecosystem disruption. Far from being a perceived crisis in the sense that DDT disrupted species reproduction, fluorocarbons have been implicated in ozone depletion, or that air pollution causes lung disease, the risks of applied genetics are still speculative.

This raises two questions. With so many well-documented hazards that fail to activate citizen interest, how do we explain the public's persistence to regulate genetic engineering without a demonstrable hazard? Secondly, what characteristic form does the regulation of biotechnology take?

To answer the first question, we have to look at the origins of the rDNA debate. It was at this point that the symbolism of genetic engineering took form, that the ethical and ecological issues captured the media's attention, and genetics policy became an issue of central interest to wider segments of society. Mixed messages were used to communicate the technology (powerful, transforming, innocuous). And finally, public awareness of genetic engineering came at a time when regulation of the chemical industry was intensified. As a new transforming technology, genetic engineering carried the burden of decades of environmental neglect and growing skepticism about high-technology solutions to ecological problems.

However, unlike the chemical industry, which has been the target of major new forms of restrictive environmental legislation, the federal government has not enacted a single new law to govern the biotechnology industry. Instead, the existing body of environmental legislation was adapted, through an incremental process, to meet the public demands for oversight. The origins of regulatory activity can be traced to the 1975 Asilomar conference where scientists made recommendations to the NIH to establish guidelines for rDNA research.

NIH's ROLE AS rDNA OVERSEER

Initially, the NIH had a clear field on the regulation of rDNA technology. The guidelines governing research were created, implemented, and modified by the agency through the RAC and the Office of Recombinant DNA Activities. The NIH had established a relatively open system of review. Except for materials designated as proprietary by private companies, all the documents, meeting notes, and public correspondence were available for public scrutiny. In addition, the RAC meetings were open to the public except when private companies sought the RAC's approval for an experimental protocol. Some viewed this as yielding too far to accommodate critical perspectives. Despite the effort by the NIH to operate in the "sunshine," there was one criticism from which it could not easily escape. The NIH was both the prime benefactor and regulator of rDNA research. The most widely cited example of what has come to be viewed as a regulatory "perversion" was the infamous role of the Atomic Energy Commission (AEC) as a key sponsor and public regulator of the nuclear industry. The NIH's dual role, however, escaped the political criticism that beleaguered the AEC and eventually caused its mitosis into two agencies, one for promotion and the second for regulation.

Other agencies of government that funded rDNA work followed the guidance of the NIH. In 1978, the Food and Drug Administration published a notice of intent to propose regulations that would require all experimental work using rDNA techniques to be done according to the NIH guidelines, otherwise the resulting product would be rejected when submitted for FDA approval. Some agencies, however, like the Occupational Safety and Health Administration (OSHA) did not initiate any new rules in the wake of the national debate over rDNA technology.

The agencies that initiated oversight activity established review panels for biotechnology. These panels frequently worked in concert with the NIH. Thus, a plant proposal sent to the RAC would receive a consultative review by the USDA before final consideration. Some informal coordination was taking place on a voluntary basis among the NIH, USDA, NSF, and FDA in the late 1970s. Since the focus was laboratory work, there was no need for a different system of containment and agency-specific safety procedures for each task. With the onset of commercial applications, however, the rationale for uniformity disappeared and, within a very short span of time, the regulatory environment became increasingly cumbersome.

There are several reasons for the sudden growth in regulatory confusion during this period. When genetic engineering was applied to product development, laboratory standards gave way to a new set of questions about safety. Each of the federal agencies already had a body of law for protecting the environment and human health. They could apply this law to genetically engineered products. But nothing fit exactly. Therefore, some agencies began to retrofit their health and safety regulations to biotechnology. There were differences in how the agencies viewed their responsibility toward the new technology. Even the definition of biotechnology was not agreed upon. When is a product a novel organism? In some instances, there was confusion over which agency would have jurisdiction over a particular product. Ordinarily, these jurisdictional problems are worked out in federal law. One commentator noted:

The foundation of the federal regulatory structure for biotechnology lies in a number of statutes . . . administered by a number of federal agencies, applied to a multitude of biotechnology products. In some cases, biotechnology products will be regulated by more than one statute while they are being manufactured, tested, and marketed. From some perspectives, such a regulatory structure must indeed appear disorganized, inefficient, and confusing.[3]

Concern over regulatory instability was a theme in a study issued by the Office of Technology Assessment titled *Commercial Biotechnology*. The central focus of the study was on the factors that determine whether the United States will have a competitive edge over other industrialized nations in the development of biotechnology. The OTA warned that "uncertainties . . . about what the regulatory requirements will be or which agencies have jurisdiction, will also affect the risk, time, and the cost of product development."[4] The report stated that jurisdictional disputes such as those between the FDA and the USDA would delay product approvals and therefore hinder U.S. competitiveness.[5]

Thus, by 1980 the industry was growing by leaps and bounds. The NIH was in a no-win situation as the surrogate regulator of the industrial sector. A new administration took aggressive measures to foster global competitiveness. And the leading environmental agency, the EPA, was looking very cautiously at taking on a new regulatory burden without an explicit congressional mandate.

BIOTECHNOLOGY ENTERS THE ENVIRONMENTAL ARENA

Two areas of conventional biotechnology had already prompted regulatory interest on the part of the EPA prior to the rDNA controversy. The first was the release of infectious wastes and the second was the use of natural bacteria and viruses as pest control agents. Under the Resource Conservation and Recovery Act of 1976 (RCRA), the EPA has authority for "regulating the treatment, storage, transportation, and disposal of hazardous wastes." This definition includes two aspects of biological wastes, namely, infectious wastes, or chemical products of biological organisms.

In December 1978, the EPA published a proposed plan for the disposal and treatment of hazardous wastes. The plan included infectious wastes as a special category of hazardous wastes. Listed in the category of infectious wastes was the classification of etiologic agents published by the Centers for Disease Control in 1974. After publication of the proposed rule, the EPA decided not to promulgate regulations for infectious waste materials. Instead, the agency chose to develop an infectious waste management plan—an information manual on state-of-the-art procedures for handling these wastes. Following President Reagan's agenda of regulatory minimalism and voluntarism, the management plan offered industry a range of options for managing the handling and disposal of infectious waste but stopped short of imposing a particular approach.

When the EPA came into existence in 1970, the federal government had already been regulating biological pesticides. Before the EPA was created, the USDA had sole authority for pesticide registration. Only a few microorganisms were accepted as pest control agents over a 30-year period. These naturally occurring organisms were registered on an ad hoc basis since, prior to 1979, there was no formal policy on the registration of biological pesticides. Table 10.1 shows the strains registered by federal agencies under authority of FIFRA through 1983.

Anticipating a rapid growth in biological control agents, in May 1979 the EPA published a policy statement on the regulation of what it termed "biorational pesticides," a category that included but went beyond microbes. Biorational pesticides were defined as "biological pest control agents and certain naturally occurring biochemicals (i.e., pheromones) which are inherently different in their mode of action from most organic and inorganic pesticide compounds currently registered with the EPA."[6]

Among the organisms covered in the policy statement were viruses, bacteria, fungi, protozoa, and algae. FIFRA also gives the EPA authority to regulate as pesticides more advanced forms of life, such as birds, insects, and aquatic

Table 10.1
Microbial Agents Registered in the United States as Pesticides, 1948–1983

Microbial Strain	Date of Registration
1. *Bacillus popilliae*	1948
2. *Bacillus lentimorbus*	1948
3. *Bacillus thuriengiensis*	1961
4. *Heliothis,* nuclear polyhedrosis virus (NPV)	1975
5. Douglas fir tussock moth NPV	1976
6. Gypsy Moth NPV	1978
7. *Agrobacterium radiobacter*	1979
8. *Nosema locustae*	1980
9. *Hirsutella thompson*	1981
10. *Bacillus thuriengiensis,* israliensis variety	1981
11. *Bacillus thuriengiensis,* aizawai variety	1981
12. *Phytophthora palmivora*	1981
13. *Colletotrichum gloeosporioides*	1982
14. *Neodiprion sertifer* NPV	1983

Source: Fred Betz, Simon Levin and M. Rogul, "Safety Aspects of Genetically-Engineered
    Microbial Pesticides," *Recombinant DNA Technical Bulletin* 6 (December 1983): 135–41.

animals. However, the agency has steered clear of regulating macroscopic control agents, exempting many of these from its rulemaking.

In 1980, the EPA drafted a 433-page document of working guidelines on the registration of biological control agents. The agency set requirements on toxicology and registration procedures as well as the assessment criteria for environmental survival of released agents, host range, and potential effects on nontarget organisms. The final version of the biorational assessment guidelines was published in November 1982.[7] This document was the centerpiece for the agency's review of the first genetically engineered pesticide, ice minus. Since the late 1940s, when the first naturally occurring microbe was registered as a pesticide, there have been 14 natural microbial agents registered in about 75 separate products. This compares to about 1,400 conventional chemical pesticide active ingredients that have been formulated into over 35,000 registered products.[8] The transition from the registration of naturally occurring microbial forms to genetically engineered variants was relatively smooth. The debate over ice minus notwithstanding, the policy foundations developed for biorational pesticides proved effective for the new generation of genetically modified strains. This was an important arena within which the EPA could test its regulatory authority over biotechnology. However, the same could not be said for microbes that came under the authority of the Toxic Substances Control Act. In this case, the only precedent for governing the release into the environment of genetically engineered organisms was chemical substances. The analogy between inert chemicals and biologicals broke down in many places. But without a new congressional mandate and with many voices calling upon the EPA to fill the void, the agency had little choice but to seek refuge in TSCA.

STATUTORY BASIS OF REGULATION

Although the EPA's direct involvement in reviewing products of biotechnology did not begin until the early 1980s, the agency's internal discussions about its role in regulating the field began soon after the first NIH guidelines were released. In 1976, the EPA's general counsel determined that parts of the Clean Water Act, the Clean Air Act, and the Toxic Substances Control Act might cover rDNA work in selected areas.

In 1977 and again in 1979, the agency's Scientific Advisory Board cited the NIH guidelines for their lack of attention to the environmental consequences of genetically engineered life-forms. Nevertheless, for about eight years the EPA was operating in the backfield, watching the NIH, Congress, and public opinion, while incrementally building its legal grounds for regulation.

The agency's early response to biotechnology was steeped in ambivalence. The EPA was gearing up for implementing major pieces of legislation enacted in the mid-1970s that dealt with toxic chemicals and hazardous waste. Biotechnology was something new that had not yet developed a broad public constituency. There were enough problems on the EPA's agenda to keep it busy for decades. Also, the heavy inflationary period in the late 1970s in conjunction with massive federal budget deficits in the post-Vietnam period forewarned of a tightening of federal spending in environmental protection. Moreover, the strongest legislative authority for regulating biotechnology was in a new and untested law, the Toxic Substances Control Act passed in 1976. Much of the EPA's ambivalence in biotechnology was centered around the role of TSCA. Under this act the EPA was given authority to develop data on potentially hazardous chemicals and to regulate chemicals both in research and commerce when they pose a threat to human health or the environment. Controversy over the applicability of TSCA to biotechnology centered around three important sections of the law.

Section 3(2) of TSCA defined a chemical as "any organic or inorganic substance of a particular molecular identity," including "(i) any combination of such substances occurring in whole or in part as a result of a chemical reaction or occurring in nature; and (ii) any element or uncombined radical." Section 4 authorizes the EPA to require testing of chemicals that may present an unreasonable risk of injury to health or the environment where there is insufficient data and experience to predict the risk. Section 5 requires premarket notification for a "significant new use" of an existing chemical substance. The applicability of these sections to microbial forms represented a formidable problem.

First and foremost in the controversy over interpretation was the question of whether microbes fall under the scope of Section 3(2). Second, there was the issue of how one determines "unreasonable risk" for a technology in which all the risks were hypothetical. And third, when a natural organism is genetically modified, under what circumstances does that constitute a "significant new use"?

Congress was considering special legislation for biotechnology in 1977.[9] Around that time, the EPA concluded it did not want to regulate biotechnology under TSCA. In testimony before the Senate Subcommittee on Science, Technology, and Space, EPA administrator Douglas Costle recommended that Congress place the statutory authority for biotechnology somewhere other than in TSCA.

I believe that effective regulation of all aspects of recombinant DNA research—including the commercial applications . . . would be better achieved by the enactment of comprehensive legislation specifically addressing itself to the unique policy issues surrounding all recombinant DNA activities.[10]

In discussing the reasons for the EPA's position, Costle outlined practical and legal considerations. TSCA was a new regulatory instrument and the EPA was having problems with its implementation for chemicals. By having to direct TSCA to biotechnology, the agency would be diverted from its central goal, namely, the regulation of toxic chemicals. Also, the agency had no expertise to launch a regulatory program in microbial products and rDNA technology. Costle added that TSCA does not fit biotechnology. If organisms are to be considered part of TSCA's mandate, then the agency might logically have to include all living things in its definition of chemical substance.

Costle also opposed trying to retrofit TSCA to biotechnology through amendments. In the spirit of the "cooperative bureaucrat" Costle conceded that in the absence of legislation "which specifically addresses the unique problems associated with recombinant DNA research," the EPA would do the best it could with TSCA and its other existing authority. The report of the Senate Subcommittee on Commerce, Science, and Technology stated unambiguously that TSCA was appropriate, although limited, for the regulation of biotechnology. There was no discussion of alternative approaches. The imperative to proceed with regulation was expressed boldly: "It is particularly important that recombinant DNA organisms intended to be released into the environment be subject to premanufacturing review and certification that they pose no significant risk."[11] Since TSCA offered the possibility of premanufacturing review, Congress leaned heavily toward it and disregarded its significant limitations.

The deficiencies of TSCA were communicated to Congress from all directions, inside and outside the government. The Natural Resources Defense Council (NRDC) wrote Senator Adlai Stevenson arguing that TSCA is not adequate to deal with toxic chemicals, and that these problems will be compounded for biological products. The NRDC pointed out that TSCA does not provide uniform regulations for all chemical substances. It could be used to regulate chemicals not already regulated by existing statutes. In this sense, it has been called "gap-filling'' legislation. Of the hundreds of chemicals that enter industrial use every year, TSCA only provides selective testing. Most chemicals will be marketed or placed in industrial use without any evaluation of their safety and environmental impact. There is no provision under TSCA that every chemical substance be tested before it is used. The NRDC underscored the point that, to act under TSCA, the EPA has to make an affirmative finding of unreasonable risk. But that would be difficult for rDNA products since adverse effects might not reveal themselves for years. Thus, if the EPA used TSCA to regulate products of biotechnology in the same fashion that it regulates chemicals, then society may be in trouble.

As it became clearer that Congress was not likely to pass new legislation, the EPA, taking note of congressional sentiment and public interest, was resolved to deal with biotechnology within its existing statutory authority. The issue of relevance was examined by the EPA's general counsel and the Administrator's Toxic Substances Advisory Committee (ATSAC), a panel created for the purpose of obtaining public input on major issues of importance to the Office of Toxic Substances.

The analysis provided by the EPA's general counsel became the legal foundation for the agency's subsequent policy on biotechnology. The general counsel concluded that life-forms are *comprised* of chemical substances, as defined under TSCA, but went beyond that—entering a realm of philosophic speculation and reductionist biology—when he claimed that life forms are *themselves* chemical substances.[12]

The EPA's legal staff invoked another argument in favor of TSCA's relevancy that was based upon Congress's intent to have the law serve in a ''gap-filling'' capacity. This was a secondary argument since one has to demonstrate relevancy before the ''gap-filling'' role becomes plausible. Congress did not intend for TSCA to be applied to a situation if there were other applicable laws. For biotechnology and TSCA, relevancy was the key issue more so than the overlapping or duplication of laws.

Between 1983 and 1984, the EPA accelerated its effort to resolve the TSCA dilemma. The first decision was whether to use the legislation; the second was how. In March 1983, the Office of Toxic Substances (OTS) prepared a draft report titled ''Regulation of Genetically Engineered Substances under TSCA.'' It concluded that a strong argument could be made for EPA jurisdiction under the law. But this draft still expressed the agency's ambivalence while hedging on the question of regulating microbes per se: ''genetically engineered microorganisms and products made from them are subject to OTS regulation *where they are used for TSCA purposes*'' (emphasis added).

The Administrator's Toxic Substances Advisory Committee also pondered the issue of TSCA and regulating biotechnology. The ATSAC issued its conclusions and recommendations after its vote in June 1983.[13]

- TSCA clearly applies to the large-scale use of nonliving biotechnology products in the environment. It is irrelevant under TSCA whether a chemical substance released was manufactured by organisms rather than by synthetic chemical procedures.

- The case for TSCA applying to the intentional release of genetically engineered living organisms is less clear. (This is consistent with previous legal interpretations inside and outside the EPA.)

- The EPA should consider the risks of the intentional release of genetically engineered living organisms just as it considers the risks from substances released by these organisms.

- Finally, ATSAC also advised the EPA to take a comprehensive approach toward regulating biotechnology, indicating that the agency should manage the potential risks from the initial production through the large-scale intentional release of genetically engineered organisms into the environment.

The final recommendation was suggestive of the "cradle-to-grave" philosophy for handling hazardous waste that was behind the EPA's approach under the Resource Conservation Recovery Act (RCRA). There is, however, considerable difference between the power of RCRA and that of TSCA for managing the environment. The former provides for much greater administrative authority and more clearly defined outcomes. Nevertheless, the EPA's ambivalence slowly shifted toward a pro-TSCA policy. Don Clay, of the Office of Pesticides and Toxic Substances, reflected the changing position in testimony before the House on June 22, 1983: "Our current view is that TSCA would provide authority to regulate recombinant DNA."[14]

Despite the broad consensus on TSCA's shortcomings for regulating biotechnology, Congress left little choice for the EPA but to proceed. In early August 1983, the EPA formally announced that it was preparing to regulate the genetic engineering industry. It added two dozen staff members and allocated several million dollars to handle new products in genetic engineering. By November of that year, Douglas Costle had signed a memorandum to (1) initiate research to evaluate the risk potential for biotechnology, (2) coordinate with other agencies on the problem of unplanned environmental releases of rDNA products, and (3) work with the NIH to ensure that future revisions of the guidelines included the EPA's concerns about the environmental consequences of rDNA activities.

With dispatch, the EPA began funding studies on the probability of escape, genetic transfer in sewage treatment, and survival of novel organisms in ecosystems. Within a year of its official announcement that it planned to regulate biotechnology under TSCA, the EPA issued guidelines to firms that planned to introduce microorganisms into commercial or industrial use.

Minimum requirements of TSCA were that firms file premanufacture notices for new chemical substances or for new uses of old substances. The EPA then examines the reported substances and decides whether there is sufficient information and, if so, whether the substance will pose an unreasonable risk to health or the environment. It is the agency's burden to demonstrate adverse health effects before it restricts the use of a product. It is the firm's burden to supply the agency with data. By the fall of 1984, the EPA distributed a draft document that outlined the type of data it required in premanufacture notices (PMNs) for

Table 10.2
The Structure of the EPA's Information Requirements for Risk Assessment of Genetically Engineered Organisms

1. Source organism
2. Alteration of source genetic material to produce engineered organisms
3. Commercial production of engineered organisms
4. Uses of engineered organisms in the environment
5. Survival, multiplication, and dissemination of engineered organisms in the environment
6. Contact of engineered organisms with other populations
7. Undesirable effects of engineered organisms (on humans, other nontarget organisms, and on ecosystems)

Source: EPA, Office of Toxic Substances, "Points to Consider in the Preparation of TSCA Premanufacture Notices for Genetically-Engineered Microorganisms," Draft document, 20 September 1984, 3.

genetically engineered microorganisms (GEMs).[15] Firms like Advanced Genetic Sciences (Oakland, CA) that had been seeking regulatory guidance from the NIH switched over to the EPA when the agency provided concrete procedures for complying under TSCA and FIFRA. As the 1990s approached, the EPA cast aside any vestiges of ambivalence about its effective statutory authority over biotechnology. As if by some act of wizardry, TSCA was transformed. "EPA believes TSCA gives the Agency considerable discretion to decide to what extent it may oversee commercial R&D field tests of microorganisms."[16]

THE EFFECTIVENESS OF TSCA: THEORY AND PRACTICE

In outlining its data requirements for risk assessment purposes, the EPA emphasized the following uses of GEMs in the environment: metal recovery, plant growth promotion, ore recovery, pollutant degradation, and field testing purposes. The agency layed out a series of sequential events (a type of fault tree analysis) that define data requirements for risk assessment (see Table 10.2). The schema defining the data requirements reflects the opinion of those advisors to the EPA who advocated a broad risk assessment that includes unintentional releases of GEMs in the production process and the exposure of workers to GEMs. Following a pragmatic path that responded to congressional interest and public opinion, the EPA put most of its resources into the ecological effects of the deliberate release of GEMs into the environment.

Public responses regarding TSCA's effectiveness for regulating the industry reflected chronic skepticism. Congress was forewarned about TSCA from both inside and outside the EPA. Don Clay of EPA's Office of Pesticides and Toxic Substances contrasted TSCA to FIFRA in testimony before Congress.

Under FIFRA, we have to have the data that we have before we will register it [a product],
and you have to have it registered before you can use it as a commercial application or
receive an experimental use permit. Under TSCA, the burden is more on the government
to prove there is a problem. TSCA is a notification scheme where you must notify the
agency 90 days in advance to manufacture a new chemical substance, which is defined
as something not already in existence or in the inventory.[17]

Independent legal experts hammered on the point that TSCA does not provide
a permit-issuing mechanism for large-scale release technologies and therefore
does not place the burden on the manufacturer to demonstrate that the product
is safe.[18]

Policy analysts at the Environmental Law Institute (ELI) summarized the
effectiveness of the EPA's PMN requirements under TSCA. According to their
report, the EPA requires relatively little health and safety testing on new chemicals prior to a PMN submission. The ELI cited as its source the Office of
Technology Assessment's review of information requirements under TSCA's
PMN program. The OTA reported that only 17 percent of PMNs have any
information about the likelihood of the substances causing cancer, birth defects,
or mutations.[19] That this record might be repeated in the EPA's review of
biotechnology was viewed with some concern among environmentalists.

Writing in the *Harvard Journal of Law and Technology*, Gary Marchant cited
problems in the implementation of TSCA. "The EPA's Office of Toxic Substances, which administers the TSCA, has traditionally been understaffed, and
there are indications that this trend is continuing for biotechnology regulation."[20]
Marchant noted that the demands on TSCA were increasing while the expenditures for regulation were being cut back.

In summary, early proposals that the EPA regulate biotechnology under authority of its existing statutes were opposed by the agency. Gradually, the EPA
modified its official position while harboring a residue of uncertainty about the
legitimacy and effectiveness of TSCA. When the EPA finally decided to regulate
the field, it chose to emphasize deliberate release experiments. Many agreed
that, of the existing statutes, TSCA had the broadest reach but lacked the muscle
for regulating GEMs. The legislation places too great a burden of proof on the
regulatory body and too little on the firm. Much would depend upon how aggressive the EPA would be in interpreting TSCA and what data and testing
requirements were promulgated and implemented for registering a biogenetic
substance.

THE SEARCH FOR A UNIFYING FRAMEWORK

The EPA's decision to regulate certain products of biotechnology was a response to some of the congressional criticism, but there still remained the problem
of jurisdiction and interagency coordination. With the existing legal basis of
regulation derived from a mosaic of laws developed for the chemical industry,

lead agencies applied the laws to different conceptions of biotechnology. For example, agencies differed on the types of genetic changes in a naturally occurring plant or microbe that constituted grounds for special risk consideration. Over a period of a decade, the NIH worked out a series of definitions and criteria for laboratory strains of GEMs. However, with the prospect of large-scale production and release into the environment, along with the entry of ecologists into the assessment process, it was inevitable that the definition of a regulated entity would be reconsidered. the NIH guidelines were useful, but not transferable to other regulatory functions.

In extending existing environmental laws to biotechnology, regulators began to recognize the gaps, inconsistencies, and potential duplication in the oversight process. For example, in some cases the USDA and the EPA had jurisdiction over the same organisms. All microbes that are plant pathogens are subject to the Plant Pest Act, and all microbes that are pesticides are subject to the Federal Insecticide, Fungicide and Rodenticide Act. Therefore, a microbe that is a plant pathogen and a pesticide is subject to both USDA and EPA laws. Without some resolution of these problems, many felt the area was ripe for internecine battles that would create delays in the regulatory process.

Industry spokespersons expressed frustration over the confounding regulatory conditions. Firms wanted explicit and unambiguous procedures for product review. They were concerned about the rise of an antibiotechnology atmosphere. The message they were giving Congress was: irrational regulation will destroy the U.S. competitive edge in biotechnology.

THE COORDINATIVE FRAMEWORK

The impetus for providing coordination for the regulatory activities in biotechnology came from the Executive Office of the President (EOP). The executive office was divided into several policy groups. One of these, the Cabinet Council on Natural Resources and the Environment, established an interagency working group in April 1984 to study and coordinate the government's regulatory policy in biotechnology. Presidential science advisor George Keyworth chaired this body, called the Domestic Policy Council Working Group on Biotechnology (hereinafter DPC Working Group). One of the principal aims of the working group was to provide "a sensible regulatory review process that will minimize the uncertainties and inefficiencies that can stifle innovation and impair the competitiveness of U.S. industry."[21] The DPC Working Group released for public comment a proposed regulatory framework for biotechnology in December 1984.[22]

The DPC Working Group maintained that the regulations in biotechnology should be consistent across agencies and within each agency. Furthermore, the regulations should be based upon the best available science. Following the Office

of Technology Assessment's appraisal in *Commercial Biotechnology*, the DPC
Working Group warned against a loss of American leadership in the science and
commercial development of biotechnology should the field be faced with excessive or irrational regulations. The operative terms were "internal harmonization," "consistency," "ease of regulatory burden," and "the U.S.'s global
competitive leadership."

The draft proposal of the DPC Working Group provided explicit indications
that the motivation of the process had more to do with stimulating economic
development than with insuring that the new technology will not pose a risk to
society.

In achieving national consistency and international harmonization, regulatory decisions
can be made in a socially responsible manner, protecting human health and the environment, allowing U.S. producers to remain competitive and, most importantly, assuring
that everyone will reap the benefits of this exciting biological revolution.[23]
Inconsistent or duplicative domestic regulation will put U.S. producers at a competitive
disadvantage.[24]

Thus, while the document itself is about regulation, its underlying premise is
to lighten the regulatory burdens of a nascent industry facing considerable public
and media attention. Despite the volume of testimony describing the limitations
of the current statutes for biotechnology and the sizable regulatory burden already
placed on the EPA, the DPC Working Group concluded that "at the present
time existing statutes seem adequate to deal with the emerging processes and
products of modern biotechnology."[25]

The DPC Working Group proposed a "science-based" regulatory program
that placed little emphasis on public participation and transcientific elements in
the risk assessment process. A scientific advisory mechanism, in the form of a
Biotechnology Science Board (BSB), popularly referred to as the "super-RAC,"
was proposed as a means of achieving scientific consensus in regulation. The
BSB was to be chartered under the Department of Health and Human Services
(HHS). The board was designed to supplement the scientific review panels within
each agency, thus establishing a two-tiered system of scientific review (Figure
10.1).

The BSB was to be comprised of distinguished scientists selected by the five
federal agencies it would serve, with two members from each agency-based
scientific advisory committee. While the BSB would review proposals submitted
to the agency committees, the rulings of the super-RAC would remain advisory.
This Office of Science and Technology Policy (OSTP) proposal of a super-RAC
within HHS drew intense criticism. The concept was significantly modified in
the final version of the coordinated framework.

RESPONSE TO THE SUPER-RAC

OSTP spent a year and a half revising the coordinated framework in response
to public comments. The strongest reaction to the document came from industry

Figure 10.1
Oversight by the Proposed Biotechnology Science Board

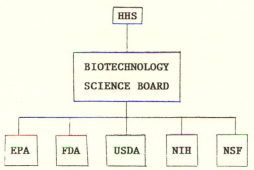

Source: Office of Science and Technology Policy, "Proposal for a Coordinated Framework
   Regulation of Biotechnology," *Federal Register* 49 (December 1984): 50856–907.

and members of the scientific and regulatory communities, while there was a
relatively mild reaction from public interest groups. OSTP reviewed 79 responses
to the draft proposal. Of those, 5 were critical of the two-tiered structure, arguing
that it was cumbersome, duplicative of the RAC, and a threat to the protection
of confidential business information. Some environmentalists supported an independent "RAC" but wanted public participation and nonagency-selected scientists serving on the committee. Their rationale was to separate the promotional
from the regulatory role of the agencies. For the EPA, this was viewed as less
of a problem than for the NIH, USDA, FDA, and NSF, since these agencies
either funded the science or promoted commercial applications of it. Many
commentators who opposed the idea of having two levels of review accepted
the idea of a second advisory panel to the individual agencies, but only upon
request of the agency.

The Facing strong opposition to the draft proposal, the OSTP gave up the two-
tiered review system. It forged a second proposal building on what it viewed as
a consensus position that coordination among the federal agencies around the
science issues of biotechnology was essential. However, the coordinating body
could not preempt the decisions of individual agencies. The OSTP replaced the
Biotechnology Science Board with a weaker body it named the Biotechnology
Science Coordinating Committee (BSCC). The new committee was chartered
by presidential advisor George Keyworth (chairman of both the Federal Coordinating Council on Science, Engineering and Technology—FCCSET, and the
DPC Working Group) on October 30, 1985.[26]

The BSCC is exclusively an interagency body, so the issue of public participation was moot. And since agencies have procedures for protecting proprietary
information, industry concerns about leaks coming from an oversight panel were
also resolved. Finally, the BSCC plays an advisory role and therefore would not
duplicate the function of the agency review panels. By situating the committee
within the executive office (under FCCSET) and not within HHS, two problems

Figure 10.2
Biotechnology Science Coordinating Committee Structure

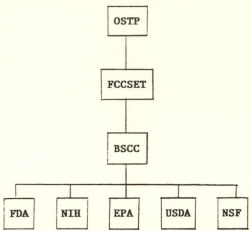

Source: Office of Science and Technology Policy, "Coordinated Framework for Regulation of
   Biotechnology," *Federal Register* 51 (June 1986): 23309–93.

were solved. First, the committee would not be viewed as being in competition
with the RAC, and second, HHS would not be seen as having a special influence
on the interagency process. Figure 10.2 shows the institutional structure of the
BSCC as finally adopted by the OSTP.

The final version of the coordinated framework for biotechnology was published in June 1986.[27] It consisted of a set of goals and responsibilities for the
BSCC, a division of agency roles for biotechnology, a set of interim policies
issued by each agency, and definitions of several key terms that OSTP hoped
would contribute to a common vernacular for each of the regulatory bodies.
Table 10.3 illustrates how jurisdictions among the federal agencies for the regulation of biotechnology products were presented in the framework.

The dilemmas over the regulation of biotechnology forced the key agencies
to communicate, if for no other reason than to resolve turf battles among themselves. Microorganisms released into the environment come under the EPA's
authority if they are pest control agents and USDA's authority if they are plant
pathogens. The framework was designed to establish a lead agency in cases
where an organism is both a plant pathogen and a pest control agent. David
Kingsbury, first chair of the seven-member BSCC, summed up three important
contributions of the coordinated framework.[28]

1. Agency jurisdiction was established in product areas, reducing misunderstanding;

2. Common definitions were proposed for uniform regulations;

3. Consistency of regulatory review was sought to preclude a situation where one agency
   accelerates the review process while another prolongs it.

Table 10.3
**Federal Agencies Responsible for Approval of Commercial Biotechnology
Products**

Biotechnology Products	Responsible Agencies
Foods/Food Additives	FDA* FSIS[1]
Human Drugs, Medical Devices & Biologics	FDA[2]
Animal Drugs	FDA[2]
Animal Biologics	APHIS[3]
Other Contained Uses	APHIS[3] FSIS[1] FDA[2]
Pesticide Microorganisms Released in the Environment	EPA[4] APHIS[3]
Other Uses of Microorganisms: Intergeneric Combination	EPA[4] APHIS[3]

* Designates lead agency where jurisdictions may overlap.
[1] FSIS, Food Safety and Inspection Service is responsible for food use.
[2] FDA is involved when in relation to food use.
[3] APHIS, Animal and Plant Inspection Service, is involved when the microorganism is plant pest,
    animal pathogen, or regulated article requiring a permit.
[4] EPA requirement will only apply to environmental release under a "significant new use rule"
    that EPA intends to propose.

Source: Office of Science and Technology Policy, "Coordinated Framework for Regulation of
    Biotechnology," *Federal Register* 51 (June 1986): 23309–93.

OBSTACLES TO A RATIONAL REGULATORY POLICY

From the early days of the RAC, semantic quagmires were commonplace in
the development of the NIH guidelines. Even the definition of "recombinant
DNA" went through several incarnations during the history of the guidelines.
The BSCC was given the task of building a common vernacular for regulation.
This task involved several difficulties at the outset. First, classification schemes
in biology are imprecise constructions. The similarities and differences among
organisms of concern for regulation may be very different from what is deemed
important for classification. Second, there are gaps in the science that may have
a bearing on the range of application of a concept for regulatory purposes. Third,
federal agencies were in disagreement over whether a product or process approach
to regulation should be adopted. The FDA was a strong proponent for the view
that products should be regulated, not processes. The agency argued that it should
not matter whether you produced a drug or a pesticide by rDNA, cell fusion,
mutagenesis, or selection. It is the safety of the product and not the method of
production that is critical. The EPA took a process-based approach, arguing that

the new genetic engineering techniques were more likely to give rise to novel strains. Initially, the EPA defined an organism as new, and therefore subject to regulation, if significant human intervention had been used in developing it. The term "significant human intervention" became the target of considerable criticism and resulted in the EPA finding a balance between a product and process approach. The product approach emphasized the traits of an organism while the process orientation selected out rDNA or other techniques for special consideration.

The concepts that needed defining were "novel organism," "pathogen," "environmental release," and "significant new use" (the latter under TSCA). The BSCC made recommendations in the coordinated framework; some definitions were generally adopted, but each agency was left to its own authority.

The BSCC constructed the elements of regulation through two concepts: "new organism" and "pathogen." A new organism is formed by the deliberate combination of genetic material from sources in different genera. These are called "intergeneric species." Pathogens were defined as any microorganism that belongs to a pathogenic species or that contains genetic material from source organisms that are pathogenic. Thus, new organisms or pathogens were subject to review under the BSCC recommendations. A controversial exclusion was added. If intergeneric DNA is used in the construction of an organism and the DNA is well-characterized (its chemical structure and function are known) and the foreign DNA contains only noncoding regulatory regions (the DNA does not produce any gene products), then the BSCC recommended that the organism should be exempt from regulatory review. The rationale behind the exemption is that if new gene products are not introduced into an organism, then nothing novel has been created. Critics responded immediately to the proposed exclusion. They argued that since regulatory genes control the production, timing, and quantities of the proteins encoded by structural genes, changing regulatory sequences could affect an organism's overall structure and ultimately its role in the environment. For the BSCC, the term "new" meant having a new phenotype.

What were the risks of genetically engineering intrageneric organisms? The BSCC agreed that the results are not wholly predictable, but it wasn't worth regulating since "the probability of any incremental hazard compared to the unmodified organism host is low." The EPA adopted the same operational definition of new organism as the BSCC but it left open the question of future regulation of intrageneric combinations and organisms developed from a single-source microorganism. By keeping these issues open, the EPA was holding on to the "process orientation" for biotechnology.

The EPA explained its emphasis on intergeneric microorganisms on the basis that (1) new traits or combinations of traits are more likely to occur through combination of genetic materials from dissimilar organisms and (2) the occurrence of a new trait or combination of traits in an organism that had not previously exhibited those traits leads to uncertainty about behavior.[29]

There were individuals within the EPA who argued that genetic engineering techniques represented a sufficient departure from natural processes that they should be regulated independently of what the source of the genes was and examined more carefully than other microbial techniques. The EPA advanced its strongest position on process regulation in its draft policy position issued in 1984 when it established its regulatory authority over "various techniques used to transfer genetic material," emphasizing rDNA and cell-fusion techniques. After a strong critical reaction, the agency modified its position by defining the relevant arena of regulation to be intergeneric microorganisms and pathogens. This meant that uses of the rDNA process on certain classes of microorganisms would be exempt from TSCA review for small-scale field experiments. In 1988 the EPA issued a new draft policy on biotechnology and eliminated the pathogen criterion for reporting under TSCA to avoid the difficulties of defining a pathogen.

The NSF, FDA, and USDA indicated that they would not adhere to the coordinated framework's definition for "intergeneric organism" or "pathogen." Meanwhile, the BSCC made an effort to define "release into the environment," which is not generally agreed upon by federal agencies.

A uniform foundation for regulating biotechnology was not achieved in the first few years of the coordinated framework. There was more complexity than the OSTP had anticipated. Each agency built a policy on its unique configuration of existing laws and those laws did not easily lend themselves to a uniform set of definitions or approaches. There were other factors operating as well. Some agencies had longer and closer associations with the industrial sector. They supported policies that eased the regulatory burden. They selected definitions that provided for broad exemptions. However, the federal government was not the only actor in this field as a new battleground developed over biotechnology in local communities and state governments. The pluralistic nature of the American system of government seems to be revitalized through jurisdictional disputes. In this case, plans to release genetically modified organisms into the environment challenged the role of state and local governments in controlling the products of genetic engineering.

LOOPHOLES, VIOLATIONS, AND CONFLICTS OF INTEREST

During the period from 1980 to 1987, when public policy for environmental biotechnology was being formed, some telltale events were reported. By themselves, as media incidents, these events tell us little. But viewed in concert and in the context of the political struggles over the role of science, they place in sharp relief the contradictions, ambiguities, and confusion that beset biotechnology during its adolescence. These events contribute to a "theater of the absurd" for this period of regulatory history. In large part, we can attribute this to the persistent confusion and ambiguity over the regulation of science and the ethical standards of entrepreneurial scientists.

The Kingsbury Affair

David Kingsbury joined the National Science Foundation as a political ap-pointee in June 1984 to serve as assistant director for biological, behavioral, and social sciences. Four years later in September 1988 he resigned his position at the NSF in the wake of a Justice Department investigation of his alleged conflicts of interest while a government employee.

At issue was whether Kingsbury was in violation of the Ethics in Government Act of 1978, which forbids senior government officials from using their public office for private gain. The act, which carries both civil and criminal penalties, requires public officials to fully disclose their financial interests. The inquiries into Kingsbury's commercial ties began after a whistle-blower from the EPA disclosed promotional material from a company called IGB Products, Ltd., a subsidiary of Porton International, that listed Kingsbury's name as a founding director. The firm's brochure stated:

IGB Products' responsibility for applied research specifically targeted at commercial developments ensures not only a forefront position for current products, but also a basis for successor products to be developed and successfully introduced into the market. Dr. David T. Kingsbury (Working Chairman of the White House panel on Biotechnology, Chairman of the Biotechnology Science Coordinating Committee of the Office of Science and Technology Policy, and Assistant Director Biological, Behavioral and Social Sciences at the United States National Science Foundation) is founding director of IGB Products and its scientific advisor.[30]

Kingsbury played a central role in the formation of regulations governing rDNA research and industrial biotechnology policy as the head of two White House groups: the Domestic Policy Council Working Group on Biotechnology and the Biotechnology Science Coordinating Committee. The question raised by Congress, the Justice Department, and the media was: Had Kingsbury been financially or managerially involved with a company that was subject to regu-lations he actively worked on while he served in the government? Had he broken any federal conflict of interest laws?

Much of the investigative journalism was done by Mark Crawford while he was with *Science*.[31] Kingsbury first maintained that he severed all relationships with the company before he joined the NSF in 1984 and that they had no authority to list him as a founding director and advisor. But evidence mounted that seemed to contradict Kingsbury's claims. There was a three-year consulting agreement dated July 24, 1985, between Kingsbury and IGB. There was evidence of pay-ments made to him of over $9,000. There was a document with Kingsbury's signature indicating that he agreed to be on the board of directors and evidence that he served on the board after 1984. First Kingsbury denied he signed a letter agreeing to direct IGB, then after seeing his signature on such a letter, acknowl-edged that he had signed it. According to *Science*, additional evidence showed that he also bought stock in the Chiron Corporation, an Emeryville, California, biotechnology company.

An internal review of the allegations by NSF personnel concluded that Kingsbury's affiliation with IGB was not in violation of the criminal conflict-of-interest statute. One of the key arguments supporting this conclusion was that the work of the BSCC involved policymaking for rDNA-related products whereas IGB's work in biologicals allegedly did not involve gene-splicing.

The Justice Department presented the case to a grand jury, which ruled the evidence did not support a criminal indictment. However, a lesser civil indictment based upon Kingsbury's incomplete financial disclosures still loomed. The final outcome was a settlement agreement between Kingsbury and the Department of Justice on the civil charges. Under the settlement agreement Kingsbury paid a $10,000 settlement award and filed amended financial disclosure reports for the years 1983–88, civil charges in connection with the Ethics in Government Act would be dropped, and no statement of guilt would be made in connection with his business dealings while employed by the government. A full public disclosure of the charges, findings of the Justice Department, an NSF review, and a congressional report has never been made.

In this instance, a whistle-blower at the EPA, a vigilant congressional committee, and an investigative journalist implicated a major figure in the development of biotechnology policy in what appeared to be a financial conflict of interest. The issue somehow faded out of sight. Kingsbury's resignation and his payment of a sizable settlement under the Ethics in Government Act, however, does not respond to the ethical and legal questions that this case raises. Are conflict-of-interest rules written differently for scientists? Had Kingsbury's reputation as a scientist provided a shield against criminal prosecution? How much did officials at the NSF know about Kingsbury's multiple roles? If he had reported all his business activities accurately, would he have been in a conflict of interest as head of the BSCC? Are the NSF and other government bodies aggressive enough in evaluating the *substance* of financial disclosures by its senior personnel?

After my service on the NIH Recombinant DNA Advisory Committee ended in 1981, I learned that some of my academic colleagues on the committee also had financial interests in the very industry for which they wrote guidelines and reviewed experiments. Financial disclosure statements were available only to the leadership of the NIH and not to other members of the RAC or the public interest community. Currently, it appears that the privacy of scientists' financial interests have primacy over government in the sunshine.

Cloning at Home

Lawrence Slot left his cancer research post at MIT to start his own firm called Genemsco. Slot believed there were home markets for the uses of recombinant DNA techniques. He developed "Dr. Cloner's Genetic Engineering Home Cloning Kit'' and began marketing it for $559. The kit provided step-by-step instructions for cloning a gene from one bacterium to another. Slot was also doing his own genetic engineering research, combining the genes of blue mollusks and bacteria to create a new building material similar to a lobster shell.

Slot set up a home laboratory in Kingston, Massachusetts. When he needed additional space, he contracted to have a basement built in his one-story home. The house was raised onto ten-foot stanchions. But one day in the summer of 1987 while the digging was underway, the house collapsed. The media reports read like a grade-D science fiction novel. ''Witnesses had claimed that hundreds of mice, rats and a 'rabbit as big as a cat' escaped the home.'' Health officials quarantined the area.[32]

After an investigation by the state and local health department, the area was declared safe. Ironically, none of the federal biotechnology statutes applied to Mr. Slot's home lab. Since he was not receiving any federal funds nor was engaged in an experiment to release rDNA organisms into the environment (although there was an accidental release), his work was unencumbered by local, state, or federal oversight. Thus, Slot, the manufacturer of home cloning kits, dramatizes how accidental releases can happen; fortunately, in his case no pathogenic or ecologically harmful agents were involved.

Vaccination and Environmental Release

In this case a congressional investigation failed to obtain a clear answer to the question of whether a scientist violated federal regulations. Science, law, and semantics came together in a most unusual way.

It started in 1981 when a biochemical virologist named Saul Kit of Baylor College of Medicine was working on a vaccine for pseudorabies. It was estimated that about 14 percent of the 70 million swine in the United States are infected with pseudorabies, which causes reduced growth, infertility, abortion, and death. By deleting a gene of an established vaccine strain, Kit developed a prototype vaccine. His research on this vaccine was funded exclusively by a company called NovaGene, Ltd., and carried out in his laboratory at Baylor College of Medicine. Field trials on the vaccine developed by Kit were conducted at a farm in central Texas in June 1984 under the auspices of Texas A&M. Approximately 1,400 swine were inoculated.[33]

About a year after the test, Jeremy Rifkin received anonymous letters (one written on NIH stationery) describing the experiment. He used the opportunity to challenge USDA for an unauthorized release of a genetically modified organism. Kit had not cleared this particular test with Baylor's institutional biosafety committee or with the NIH. He didn't believe he had to. However, the test was done with the knowledge and approval of the Veterinary Services Division of the USDA.

Kit was caught in a regulatory ''Catch 22.'' He was trying to comply with one agency that had one set of criteria while another agency had a different set of criteria. Two questions kept reappearing in the investigation. Was the vaccine a recombinant organism? Is the vaccination of 1,400 pigs considered an environmental release?

At the time of the field trials USDA did not consider the vaccine to be a recombinant organism, which made the question of environmental release moot. However, by November 1985, the USDA had declared the vaccine a recombinant organism. USDA officials claimed that if the decision had come earlier it probably would not have changed the agency's role in the vaccination experiments since it would not have prompted a special review from USDA's Agricultural Recombinant DNA Review Committee.[34]

Kit told a congressional committee that his vaccine was not a recombinant virus since it doesn't have DNA from two species mixed together and wasn't constructed in a test tube. The conventional definition of an rDNA organism is one created by combining the DNA of two taxonomic species *in vitro*, resulting in an organism that can replicate. Kit claimed he used the strain of the virus that already has the deletion and only used rDNA techniques to learn that the deletion existed.

This is where hair-splitting Talmudic scholarship was required. Kit claimed: "I regard the material as recombinant DNA derived, but I do not regard the virus as a recombinant."[35] It was this language that confounded members of Congress and the regulators as well. But the Baylor College IBC did not agree with his interpretation. "The final product as the vaccine *per-se* must be regarded as a recombinant DNA virus or as containing recombinant DNA molecules."[36]

Regardless of whether the organism is defined as a recombinant or not, Kit maintained that a vaccination cannot be considered an environmental release. He brought to his defense a USDA Veterinary Services Memorandum that stated: "In normal husbandry and laboratory practices, veterinary biological products are not considered to be released into the environment."[37]

On the other hand, according to guidelines issued by the Department of Health and Human Services, NIH and IBC approval was required for any experiments that result in the "deliberate release into the environment of any organism containing recombinant DNA, except certain plants." To complicate matters further, the experiments Kit did in the laboratory were exempt from the NIH guidelines, but the release experiments were not. Thus, Kit was caught in a regulatory imbroglio where there was overlapping and conflicting authority, obscure science, and inconsistent criteria. This was the kind of confusion that benefited no one and demanded a coherent, integrated system of regulation with consistent definitions and an unambiguous process for dealing with unusual cases.

Trees and Containment

A second controversy over deliberate release involved a plant pathologist at Montana State University named Gary Strobel. Between 1983 and 1984 Strobel injected genetically altered bacteria (Rhizobium) into American elm trees a few miles from the campus. Similar tests were done in South Dakota, Nebraska, and California. Officials at the EPA learned about Strobel's tests in the aftermath of an investigation that began in the summer of 1987. Strobel inoculated trees on June 18 of that year, around the same time he sent the EPA the request for review of the small-scale field test. He violated the EPA's regulations requiring 90-day prior notification.[38]

Strobel maintained that the federal rules were inconsistent, imprecise, and confusing. He characterized his flouting of the rules as an act of "civil disobedience."[39] Officials at his institution, who feared loss of federal funds, had a more sober reaction.

Strobel's elm tree field tests are reminiscent of Advanced Genetic Sciences' rooftop arbor experiments. Both AGS and Strobel viewed the tree as a closed system and not a conduit for an environmental release. The thing that distinguished this case from Kit's pseudorabies release is that Strobel openly flaunted the regulations as irrational and repressive to free inquiry. Scientists had never overtly used a metaphor like civil disobedience to describe their opposition to regulation. Strobel was eventually reprimanded by the EPA for his unauthorized release of genetically altered bacteria. The agency also informed Strobel that for a period of one year any proposal he might have for field testing that is subject to EPA's policy must be cosponsored by the university or some other responsible party and be approved by his university's Institutional Biosafety Committee.

These are some of the strange occurrences during the early period of commercialization in applied genetics when regulations were in flux, controversies had short half-lives, the cynicism of scientists toward government oversight was widespread, and the boundaries between science and commerce in biology had disappeared.

Biotechnology policy became an irritant and an embarrassment to the Bush Administration because the effort to harmonize regulations among different agencies of government by the coordinating function of the BSCC resulted in a deadlock. Basic philosophical diferrences toward regulation prevented the BSCC from reaching a consensus. The failure to resolve the issue of the proper scope of oversight for biotechnology was an important source of instability for the new industry. With the charter of the BSCC due to expire in 1990, the Bush administration used this opportunity to create a consensus position at another administrative level. In place of the BSCC, a similar interagency body called the Biotechnology Research Subcommittee (BRS) was established. This redefined interagency committee was given the task of providing scientific back-up for policies established by the White House Council on Competitiveness.[40] At this stage, the oversight of biotechnology was directed by those fully invested in commercial development.

11

Biotechnology Assessment:
Dilemmas and Opportunities

Before the introduction of a new biotechnological product or licensing of a new biotechnological production plant, its impact on the general welfare, health, economy, labour situation, culture and socioeconomic structures, etc. should be studied.

Cary Fowler et al., 1988, Rural Advancement Fund International[1]

Biotechnology is a global issue. It cannot be assigned such attributes as positive, negative, or neutral. Like any other technology, it is inextricably linked to the society in which it is created and used, and will be as socially just or unjust as its milieu. . . . rational biotechnology policy must be geared to meet the real needs of the majority of the world's people and the creation of more equitable and self-reliant societies while in harmony with the environment.

The Bogeve Declaration, 1987[2]

Previous chapters in this book have shown how the industrialization of applied genetics has contributed to a new generation of social, ethical, legal, and ecological problems. The R&D and industrial sectors in biotechnology have aggressively sought product opportunities in the tradition of other high-tech ventures like microelectronics, computers, and robotics. But these industrial revolutions cannot compare to the commercialization of genetics in the public apprehension associated with their successes. Geneticist Steve Gendel asks: "Why has biotechnology become such a focus for ethical, social, and economic debate while other technologies are all but ignored?"[3] His answer focuses on the subject matter. "Clearly biological issues touch a sensitive aspect of our culture and leads to deeper and more passionate examination of issues than do issues raised by any other technology."[4] I would argue that part of the difference lies in the fact that concurrent with the genetics revolution there have been challenges to the traditional ways of addressing the externalities of industrialization. These challenges are confounding to government regulators and entrepreneurs who place their confidence in the established norms of social governance.

Industry and the regulatory sector share a commitment to incrementalism, although, by virtue of new legislation, the latter may be thrust into an alternative mode of handling the potentially adverse consequences of technological innovation. Established environmental laws and ethical traditions applied to biotechnology are being stretched to their limits. The stresses have rekindled ideological debates on the responsibility of government toward new technologies. In this chapter I review some representative responses to the social management of technology. Then I explore the role technology assessment has played in evaluating a few high-profile products arising out of the new biotechnology industry. Finally, I outline a proposal for a more highly integrated approach to the review of new technologies that takes us beyond hazard control to social governance of innovation.

POLITICAL IDEOLOGY AND BIOTECHNOLOGY

Environmental Traditionalists. Environmentalism, as distinguished from political and social ecology, is rooted in the constellation of laws that protect humans and segments of the ecosystem from the products and processes of industrialization. The vast majority of these laws that have been enacted at the federal level came in response to public concerns over the hazards of the chemical, nuclear, and fossil fuel industries. Environmental traditionalists advocate a modification of the current regulatory system to address the problems of biotechnology. Some modifications, additions, and adaptations to the established regulatory regime of FIFRA, TSCA, and to a lesser degree the Food and Drug Acts, have already been made in response to biotechnology. The vast body of environmental law has not been amended by Congress. However, minor modifications of the existing statutory framework are well within the purview of the traditionalist response to the biotechnology revolution.

Reactionism. Among those who reject environmental traditionalism are individuals who advocate a libertarian model of technological innovation. According to this view, society should not assume the technology is hazardous before it is *proven* hazardous. Secondly, it is argued that the costs of pursuing "phantom hazards" is too great for society to bear. They cite ice minus as an example. It took five years and millions of dollars of regulatory review and litigation before an outdoor field test was permitted for an organism with a "mere" single gene deletion. The tradition of reactionism has attracted those who would eliminate the Delaney amendment for food additives, do risk-benefit balancing in assessing technological hazards, and place more emphasis on tort law and less on regulatory bureaucracy.

Social Ownership. Proponents of social ownership or social directorship of biotechnology argue their case from either a capitalist or socialist perspective. From the capitalist perspective, social investment should reap social benefit, while private investment should reap private benefit. Since the entire field of biotechnology arose directly from federal funding of molecular biology, under the logic of the economic system the public sector should be a key beneficiary in the outcome. In support of this view Barry Commoner stated: "We have to ask ourselves about the morality of allowing publicly produced knowledge to be taken over by the owners of capital."[5] This view is antithetical to the patenting of life-forms or the private appropriation of federally supported discoveries.[6]

From a socialist perspective, society will get the most out of biotechnology if its productive resources are directed by a state planning group or decentralized planning councils representing broad constituencies in society. Proponents of social ownership cite the direction that biotechnology takes under free market conditions. Profitability, and not social needs, dictates product development.

Commoner, who advanced a similar argument for the direction of the energy industries,[7] cited public control of technology at the sources of innovation and production as the solution. "A fundamental question that any of us concerned with biotechnology have to deal with is the problem of governing the development of a new industry. I'm not talking about regulating its impact on the environment. I am talking about the social governance of the means of production."[8]

Without socially directed industrial development, Commoner and others argue, biotechnology will serve the interests of large established industrial corporations (petrochemical and agribusiness) and leave to pure chance the match between the productive capacity of the new technology and its contributions to the central problems of civilization (malnutrition, disease, environmental degradation, lack of inexpensive and clean sources of energy, prohibitively expensive health care).

A FOURTH WAY: MARKET INNOVATION AND SOCIAL SELECTION

Socialist solutions to the problems of postindustrial capitalism have lost much of their currency since the Reagan-Gorbachev era. With the world's major socialist economies (China and the USSR) exploring market alternatives, the rhetoric of centralized planning has far less appeal, even among democratic socialists. There is still much to be socialist about beyond the command economy and state ownership of the modes of production, particularly the public's role in determining the size and allocation of the federal budget for social needs. But state economic socialism does not provide a sensible solution to harnessing biotechnology for the masses—at least not in the advanced capitalist nations.

What alternatives are there beyond the three cited for the governance of biotechnology? I shall describe a system of social guidance that I refer to as "market innovation-social selection." It is based on five premises.

1. The innovation sector and the social guidance sector shall be distinct. The main purpose of the former is to create new marketable ideas—to always be innovating—while the latter must evaluate these ideas within a highly articulated system of social directives.

2. The state shall expand its role in the assessment of new technologies. All new technologies must be evaluated on health and safety, ecological, equity, and ethical criteria.

3. Public participation in the assessment of new technologies shall involve all levels of political jurisdiction.

4. The state shall support maximum innovation in the private sector, but by a conscious process of selection, reinforce those innovations that meet important social needs and provide selective negative pressures against unneeded or unwanted innovations.

5. Only in cases where a robust system of private initiatives fails to meet public needs shall the state assume the role of innovator. However, in such cases (e.g., orphan drugs or recycling projects), innovation and social governance shall be the function of independent government bodies.

This system of social guidance for technology is modeled on Darwinian principles where two opposing processes (mutation and selection) provide the basis of growth, change, and balance. Innovation is essential for technological change. But the state's role in selecting among competing technologies has been too limited and weak, and leaves too much to the control and self-interest of the innovation (production) sector. The current system is too product-centered. As a consequence it fails to account for technological directions. Social choices about the broad goals of technology are often the result of, or held hostage to, microeconomic decisions. The position I am advocating builds on a nascent form of technology assessment that began nearly two decades ago.

CONTRIBUTIONS TO TECHNOLOGY ASSESSMENT

The Office of Technology Assessment (OTA) was created by an act of Congress in 1972. Under its congressional mandate, OTA's principal function is "to help legislative policymakers anticipate and plan for the consequences of technological changes and to examine the many ways, expected and unexpected, in which technology affects people's lives." Since its inception, OTA has published numerous studies and briefing papers on topics ranging from "Star Wars" to toxic wastes. In the arena of biotechnology, OTA has produced ten major studies and four background papers (see Table 11.1).

Despite the impressive volume of OTA studies in biotechnology, no generic framework of technology assessment links these diverse studies. Each assessment is carried out by a different staff, including a new project manager. Individuals who lead the studies are brought to the OTA for a few years and usually leave after their project has been completed. Topics for OTA analysis are proposed by congressional leaders such as the chairperson of a committee or subcommittee.

The OTA has developed its approach to technology assessment from a literature that has weathered 15 years of critical discussions. The end product for the OTA is a document that provides policy alternatives to Congress. For each study, the OTA convenes an advisory panel. The composition of the panel can affect how the assessment is carried out, which topics are emphasized, and which issues

Table 11.1
Studies in Biotechnology Produced by the Congressional Office of Technology Assessment, 1981–1989

Title	Date
1. Impacts of Applied Genetics: Microorganisms Plants and Animals	Apr. 1981
2. The Role of Genetic Testing in the Prevention of Occupational Disease	Apr. 1983
3. Commercial Biotechnology: An International Analysis	Jan. 1984
4. Human Gene Therapy: Background Paper	Dec. 1984
5. The Status of Biomedical Research and Related Technology for Tropical Diseases	Sept. 1986
6. Ownership of Human Tissue Cells	Mar. 1987
7. Public Perception of Biotechnology	May 1987
8. Commercial Development of Tests for Human Genetic Disorders—Staff Paper	Feb. 1988
9. Transgenic Animals—Staff Paper	Feb. 1988
10. Federal Regulation and Animal Patents— Staff Paper	Feb. 1988
11. Mapping Our Genes	Apr. 1988
12. Field-Testing Engineered Organisms: Genetic and Ecological Issues	May 1988
13. U.S. Investment in Biotechnology	July 1988
14. Patenting Life	Apr. 1989

are muted. A close examination of the studies reveals that technology assessment means many things to the OTA. To illustrate, in its study titled *Commercial Biotechnology*, the OTA placed considerable emphasis on what the federal government can or should do to protect America's competitive position in biotechnology.[9] When the issue of environmental release of genetically modified organisms evoked public attention, a follow-up study gave primary attention to the ecological consequences of releasing novel organisms.[10]

In applied genetics, the frameworks for technology assessment employed by governmental bodies often do not reflect the full scope of public discourse and critical debate. Primary emphasis is given to environmental and health impacts and/or to economic consequences both adverse and positive. This traditional orientation is grounded in a neoclassical paradigm that is comprised of several elements.

First, the private sector is primarily responsible for technological innovation. Second, the market system determines the social utility of the innovation. Third, the governmental sector protects society from the negative externalities of technology on human health or the environment. More recently a fourth component was added, and that is government's role in stimulating and protecting innovation in U.S. science and industry.

CRITICAL SCHOOL OF TECHNOLOGY ASSESSMENT

Starting in the mid-1970s, a new critical school of technology assessment began to gain support, drawing its theoretical underpinnings from Lewis Mumford's humanistic sociology, decentrists like E. F. Schumacher, neo-anarchists like Murray Bookchin, and "soft energy" advocates like Amory Lovins. This movement rejected the neoclassical model of assessment and developed an alternative set of criteria against which to evaluate technology.

Factors like scale of technology, job displacement, ecological value, dependency on nonrenewable resources, quality of work, and degree to which the technology can be democratically managed became the new elements of an assessment program. The concepts of efficiency and externalities that represented the marrow of the traditional paradigm were replaced with norms that encompassed broad social and cultural categories.

By recasting the parameters of technology assessment, critics of postmodern industrial capitalism expressed an allegiance to a new set of social values, such as deep ecology, the humanism of work, and the stewardship of nature.[11] Technology became a vehicle through which to raise questions about the composition, control, and direction of America's industrial economy. Environmental interests merged with the interests of democratic decentrists and those who sought a stronger role for labor in setting economic policies.

Biotechnology is notable in that it is the first modern technological revolution that can be evaluated, in advance of its industrial deployment, by the technology assessment framework emerging from the new critical school. Despite this opportunity, little has been done to widen the lens of analysis for the first generation of biotechnology products.

When public issues around biotechnology are debated within the traditional assessment framework, much of the public discourse is left freestanding, without an institutional anchor. In some instances, issues raised by the public are viewed as unrealistic and dismissed as being naively romantic. To explore this point further I introduce seven factors for discussing new technologies: ecological impacts, human health effects, ethical soundness, economic productivity, distributive justice, social need, and market demand (see Table 11.2).

A product or new technology may thus be classified according to seven variables. This particular set of variables is useful in examining the current controversies over biotechnology. I shall use a ternary value system to distinguish different weightings of these variables: positive ($+$), mixed or uncertain (0), and negative ($-$). With a total of seven variables, each with one of three values (0, $+$, $-$), the possibilities are 3^7 or 2,187 cases. By assigning one of the three

Table 11.2
Relevant Factors in Technology Assessment

1. Ecological Impacts (EI): direct and indirect effects on natural systems, sustainable utilization of resources, species, and habitat diversity.
2. Health Effects on Humans (HE): physical and psychological effects and their distribution in the population.
3. Ethical Soundness (ES): impact on ethical norms, moral beliefs, or general quality of life principles.
4. Economic Productivity (EP): contributions to efficiency of production, new wealth, international markets, or new jobs.
5. Distributive Justice (DJ): contributions to the distribution of existing resources or the concentration of new or old wealth.
6. Social Needs (SN): contributions to essential needs related to health, well being, education, social order, or environmental quality.
7. Market Demand (MD): fulfillment of an existing or potential market demand sufficient to justify commerical development whether or not it meets a social need.

Table 11.3
Technology Assessment Index

Index Variable	Best Case	Favored	Mixed	Worst Case
EI	+	0	−	−
HE	+	+	0	−
ES	+	+	0	−
EP	+	0	+	−
DJ	+	+	−	−
SN	+	+	−	−
MD	+	+	+	−

values to each of the variables, we can generate a specific assessment profile. Table 11.3 illustrates four cases—a best, favored, mixed, and worst assessment for a hypothetical technology.

The distinction between social needs and desires provides a key demarcation between market and nonmarket approaches to technology assessment. In the former, desires are measured by consumer or market demand and needs are a function of desires. In other words, people need what they choose in the marketplace, and what they do not choose they do not need. There are, of course, always exceptions to the general rule. Poor people cannot fully express their needs in the marketplace. Therefore, they have needs that cannot be measured by consumer choice.

The critical school of technology assessment rejects the compression of needs and desires. It is acknowledged instead that something can be highly desired but not needed, or greatly needed and not desired. The orthodoxy of neoclassical economists, is blinded to the problem that wants are sometimes created by the production sector through advertising.

The fact that a firm develops and successfully markets a product is not *prima facie* evidence that the product meets an overall social need. Otherwise the following statement would be a tautology: any successfully marketed product produces an abundance of social utility over social disutility. The *informed* public response to a product or technological innovation provides a clearer indication of social needs than whether the innovation is efficient, profitable, or marketable.

In addition to social needs and market demand, the assessment criteria contain two biological factors (Ecological Impacts and Health Effects), two normative factors (Ethical Soundness and Distributive Justice), and an attribute that drives innovation and includes the profit element (Economic Productivity). The criterion of ethical soundness is an indicator of whether the product or technology is compatible or conflicts with existing ethical norms. Distributive justice refers to the impacts of the product or technology on the distribution of property/ wealth. A negative consequence indicates a redistribution of resources in favor of the "best off" in society. The normative dilemmas of technologies are addressed in a variety of ways, sometimes within clear institutional parameters and sometimes outside of those parameters. As examples, ethical problems associated with the application of human gene therapy are the responsibility of the NIH's Recombinant DNA Advisory Committee. Alternatively, ethical questions associated with creating transgenic animals have no natural venue within the federal regulatory system.

Assessing technologies on the basis of their human health and ecological impacts rarely is an uncomplicated matter. Knowledge is often incomplete. Predictive models are seldom dependable. Long-term effects are always problematic. Biotechnology, however, stands apart from other technological developments in one important respect: many of the questions about the impacts have been raised at the early stages of product development. This is characteristic of a social order that is responding to historical learning. Local communities that have questioned different facets of biotechnology responded out of their own experiences with technological risk. But by raising the health and ecological safety issues early in the maturation of a technological revolution, the uncertainties are even greater. In some cases the risks can barely be articulated. The reaction to a new technology may result from a mistrust in technology per se, or there may be something about a particular technology or its product that strikes a raw nerve in a community. The novelty and unfamiliarity of biotechnology is certainly one factor that contributes to the public response. But that is only part of the equation. Some of the issues go much deeper to questions of control and accountability.

Traditional technology assessment has focused on the externalities of what are deemed socially useful products or processes. In theory at least, the constellation of environmental laws passed in the 1970s was designed to protect human health and natural resources. Among the most desirable and least controversial technologies are those considered socially useful (needed), sound on human health and ecological grounds, either ethically positive or at least ethically neutral, and favorable on considerations of equity. The "favored" column in Table 11.3 might represent the development of a new vaccine to treat an infectious disease. The zero values of EP and EI indicate that the drug in question is not a big profit item and has no important ecological impacts (positive or negative).

Many public debates over technology arise because of the uncertainty inherent in explicating the health and environmental effects, regardless of the social utility. The early recombinant DNA debate serves as a useful example. In the mid-1970s, most scientists did not question the social utility of recombinant DNA molecule technology. But they did question the safety of certain classes of experiments. An argument frequently heard is that society must always take risks if it wishes to support technological progress. Only a few scientists questioned the premise that genetic engineering meant progress, or the extravagant claims about progress.[12]

During this early period, the scientific community was roughly divided into two camps, a division marked by the following query: Do we have sufficient information to insure confidence in the safety of transplanting genes from one species to another? The public debates that ensued did not accept those boundary conditions. They cited the ethical issues of human genetic engineering and the threat of biological weapons. Some individuals argued that we do not need rDNA technology or that there are moral reasons not to develop it. However, in the early stages of the rDNA controversy, the social utility of gene-splicing was not seriously questioned by more than a few faint voices in society.

In my criteria for technology assessment, I draw a distinction between a technology that is not needed and one that is not desired. To say that a technology or one of its products is not needed is to assert either that (1) we have an adequate substitute for it, (2) it is inconsequential (a marginal addition to the quality of life), or (3) it is detrimental to the physical, psychological, or social well-being of individuals.

A technology is undesired by some constituency when it is perceived to offer a greater balance of negative to positive utility. The public responds to undesired technologies exclusively through the marketplace. As an example, suppose a new technology is developed for sex selection of children. It may be argued that this technology is not needed by society (there are no sound reasons for selecting the sex of a child) and that it is also unethical as it may create imbalances in the world population or reinforce misogynic social mores. But this argument will not convince everyone and there will most assuredly be a demand for sex selection if it is available.[13] The "mixed" column in Table 11.3 illustrates this scenario. Alternatively, there are technologies that some experts believe society needs but popular opinion is against, such as nuclear power. For commercial genetics, the social discussions over technology have become increasingly complex. In some instances, debates are fruitless because proponents construct basically incommensurate arguments derived from the different variables for technology assessment. A characteristic of such debates is that claims and counterclaims fall on unreceptive ears. There are ideological filters within each camp that treat information or analysis derived from the other as illegitimate. I shall illustrate these along with other issues of technology assessment by applying the assessment parameters in Table 11.2 to several early and promising products of biotechnology. The first case I shall consider is bovine growth hormone (BGH).

BOVINE GROWTH HORMONE

Pharmaceuticals and veterinary drugs represent two of the largest markets for biotechnology. One of the earliest products developed as an animal drug is bovine growth hormone (BGH), alternatively called Bovine Somatotropin (BST). This hormone is responsible for regulating the volume of milk production in a cow. When injected into dairy cows at the rate of 25–50 milligrams per day, the hormone has resulted in significant increases in milk production. Preliminary studies have indicated a 10 to 40 percent increase in milk yield. A Monsanto Corporation official described the economic efficiency of BGH in the following terms: "In the future, a farmer using BST will be able to produce as much milk with 70 or 80 cows as can be produced with 100 cows today, use 15 percent less feed to produce that milk, and finally have a chance to be more profitable."[14]

Media reports of bovine growth hormone indicate mixed responses from the public for the seven categories in the technology assessment index. In regard to social utility, we can ask whether the U.S. economy or the consumer requires higher milk yields from the U.S. dairy herd. There is evidence to the contrary. In 1986, milk production was sufficiently high to impel the government to buy and slaughter dairy cows. In that same year, New Jersey milk producers initially were blocked from selling milk to New York City residents at prices below their New York counterparts. Subsequently, the Jersey milk did enter the New York market and caused a slight decline in consumer milk prices. Thus, if BGH's most important contribution to social utility is expanding milk stocks, then it is a good example of a product that, on the face of it, people do not need. This argument was made in a report by the Rural Advancement Fund International.

The development of new technology to dramatically increase milk production comes at a time when the United States is already plagued by massive dairy surpluses. The US Department of Agriculture's price support policies have led to a government-owned stockpile of more than 3 billion pounds of dried milk and cheese and a federal dairy program which has cost more than $1 billion annually in recent years. In April, 1986, the government launched a $1.8 billion surplus reduction programme which pays farmers to slaughter their dairy cows or sell them for export.[15]

Another analyst noted that "bovine somatotropin is a technology that is entering an industry that is experiencing excess resource utilization, overproduction, and an excess commitment of land, labor and capital."[16]

Early public and media reactions to BGH were guarded. There were no probing investigations into the potential human health or ecological effects, but questions were being raised. Will the quality of milk derived through BGH treatment be adversely affected? Could BGH injected into cows produce by-products in the milk that might prove injurious to consumers? Samuel Epstein, professor of occupational and environmental health at the University of Illinois Medical Center, reviewed the test data and claims associated with BGH. He concluded that the regulatory agencies and BGH producers have failed to demonstrate either safety or efficacy. Epstein stated that "the use of milk hormones poses serious risks of adverse public health effects that have not been adequately considered, in spite of continued unfounded but strident industry and industry contractee assurances of safety."[17]

If we assume the BGH milk product is completely safe but of inferior quality, on the face of it this can be interpreted as socially undesirable. However, if the public is given a slightly inferior product but at a lower cost, this might be interpreted as providing a mixed social utility for BGH. The technological debate might then take the form of questioning the tradeoffs and the decision process. The tradeoffs might involve striking a balance between lower nutritional quality and reduced cost or higher production efficiency? However, milk, unlike gasoline, a product for which the consumer chooses between price and octane, does not lend itself to such tradeoffs. It is not likely that the public will tolerate a bifurcated market with an inferior quality of milk.

To date, FDA has issued assurances that milk produced by cows receiving BGH poses no risks to humans and that the milk is indistinguishable from that of untreated cows. Nevertheless, a consumer boycott supported by the Foundation on Economic Trends showed some signs of success. On August 23, 1989, five of the nation's largest supermarket companies declared that they would not carry dairy products that contain milk from cows treated with bovine growth hormone.[18]

Beyond the issue of the quality of milk, other critics questioned the social and economic value of BGH. A 1984 report from a group of Cornell University agricultural economists headed by Robert Kalter stated that the use of BGH in conjunction with other technical changes will result in a reduction in the number of producing cows and eventually in the number of dairy farms in the United States.[19] The increasing marginalization of small farmers in the United States has been a cause célèbre for certain public interest sectors, like the Wisconsin Family Farm Defense Fund and the Rural Advancement Fund International (RAFI), which have lobbied to protect the small family farm and the unique culture it represents. The Kalter study prompted a new wave of social criticism against a showcase product of biotechnology.

Jack Doyle, who directs the biotechnology project for Environmental Action, wrote in his book *Altered Harvest* that BGH was part of a trend toward the economic concentration of agriculture through genetic engineering.[20] The same issue was addressed by Jeremy Rifkin, who petitioned the Food and Drug Administration to block approval of BGH. In his petition, Rifkin argued that "entire dairy communities could well be economically and socially devastated by the widespread commercial use" of the hormone. Rifkin also cited the overproduction of milk as a social disutility.[21] Some agricultural economists believe that overproduction resulting from BGH may force an end to price supports in the dairy industry, which could adversely affect small operators.

In response to these criticisms, FDA commissioner Frank Young consistently maintained that, beyond health and nutritional issues, FDA has no responsibility for the social utility of the product. In most instances, there is no context within the regulatory agencies to raise and debate such issues.

Finally, we turn to the ethical consequences of BGH. Concerns were raised about how the use of the hormone will affect the treatment of animals. Even though milk cows are imprisoned "beasts of burden" that have been continually adapted to high-technology methods of milk production, many would still afford the animals some species considerations. Protection against the needless infliction of pain and suffering is one example of a widely respected right for animals. Rifkin's petition to the FDA on bovine growth hormone was cosigned by the Humane Society of America and the Wisconsin Family Farm Defense Fund. It stated that the use of BGH on cows will make them more subject to stress and disease.

Treatment with this hormone [BGH] will mean that dairy cows will be under even greater production pressure. Their body metabolism, which is already being pushed to the limits on factory farms, will be stressed further. Current intensive dairy husbandry practices are responsible for a host of so-called production-related diseases which result in stress and suffering to dairy cows that, as a consequence, "burn-out" in 3–4 years.[22]

Michael W. Fox of the Humane Society advanced the ethical argument in a published letter in *Science* in which he postulated the effects on cows of administering a growth hormone that hyperstimulates them to produce 20–40 percent more milk.

Under the present intensive husbandry conditions, the average dairy cow is spent by the time it is 4 to 5 years of age because of so-called production-related diseases. It is highly probable on the large dairy farm that hormone-stimulated cattle will burn out at an even faster rate, hence the concern that this treatment will increase their suffering as well as the incidence and severity of production-related diseases.[23]

In Fox's view, not only is BGH a product society does not need but it is one that fosters further abuse to farm animals. David Kronfield at the University of Pennsylvania predicted that, at high levels, BGH will result in reduced reproductive efficiency, mastitis, decreased immune function, and many other diseases associated with early lactation. However, when BGH is used at low levels in conjunction with good animal husbandry, the animals may not experience ill health.[24]

The animal rights movement has gained considerable public support over the past decade; however, it has had little effect on the treatment of farm animals. Much of the public's attention has focused on experimental and fur-bearing animals as well as endangered species. Except for hard-core animal liberationists, moral vegetarians, and the radical fringe of animal protectionists, public response to the unethical treatment of food animals has been relatively unimpressive. More significantly, there are no institutional frameworks that provide a context for critical discussion of this type of technological impact. Laws and regulations offer a simple response to the ethical argument: it is not within our mandate to evaluate the feelings and/or quality of life of food animals per se.

The case of bovine growth hormone illustrates the diversity of the issues around a single product of biotechnology. Applying the technology assessment criteria and assuming a best-case scenario for BGH we have the following values.

Index Variable	BGH Assessment Values
EI	0 or −
HE	0
ES	0
EP	+
DJ	−
SN	−
MD	0 or −

In other words, this is a product that the consumer doesn't need, whose demand is primarily with the large dairy farmers, that currently tests neutral on health and ecological impacts, and is ethically suspect, but whose economic efficiency has been demonstrated in pilot studies. This issue is characteristic of a class of social concerns that have no natural place in the public policy arena. Without a well-defined institutional setting or a specific body of relevant law, these issues remain free-floating and often unresolved. Once BGH passes the safety and efficacy tests of veterinary drugs, existing statutes have little to offer opponents. Their fallback position calls for labeling of BGH milk and letting consumers make the choice among technologies.

HERBICIDE-RESISTANT PLANTS

Among the earliest research programs in the agricultural applications of genetic engineering is a plan to increase crop productivity by reducing weed interference during different stages in the growth cycle. Since World War II, pesticide development has been the leading factor in improving efficiency of agricultural production. The selection of robust seed strains and the development of new hybrid varieties have also contributed to improving output per acre. Herbicides, directed at the early stages of the seeding process, played a lesser role in agriculture compared to insecticides until the early 1970s when their use rose dramatically. Currently, herbicides constitute about 60 percent of the total tonnage of pesticides used in United States agriculture.

Herbicide use got a boost during the Vietnam War in the late 1960s and early 1970s when it became an instrument of the military for destroying the dense foliage in strategic areas of Southeast Asia.[25] Herbicides are also widely used in the United States to clear electric lines and railroad rights-of-way of obstructive plant growth, and for home lawn care.

Once seedlings are formed, many herbicides are not selective between weeds and the crop itself. Recombinant DNA techniques gave scientists a new tool to develop herbicide-resistant plants. Three justifications were advanced for this product. First, in areas where there is crop rotation the residue of herbicides used for one crop may damage the second crop. The same reasoning can be applied in cases where there are multiple plantings without crop rotation. The second planting can be affected by herbicide residues from the first. If herbicides are used to deweed an area prior to seeding, chemical residues could damage the new crop during its early stages of development. One solution to this problem is to use herbicides that degrade quickly; another is to replace herbicide use with labor-intensive forms of weed control. A third solution is to make the crop herbicide-resistant.

The second justification for developing herbicide-resistant plants is that chemical weed control could then be carried out at every stage in the crop's development. With the first justification, the overall use of herbicides need not increase. However, the second justification provides a very different picture of chemical agriculture. It suggests that herbicides will be used in multiple stages during crop development or on crops currently considered far too sensitive for such use. A third incentive for using herbicides is that it can substitute for turning over or tilling the soil after each planting. No-till farming not only saves on labor but also reduces the erosion of nutrient-rich topsoil.

A selection from Calgene's 1984 annual report to stockholders reveals how the use of herbicides is being viewed by the R&D and investment community.

Herbicides are used to kill weeds that reduce crop yields. The worldwide market for herbicides is $4.5 billion. Historically, this business has been characterized by high growth, high margins, and strong proprietary positions. In recent years, however, markets have matured and margins have eroded as a result of the introduction of new competing chemicals, intense price competition, and expiration of patents. Also, the use of chemical herbicides has become a highly visible environmental and regulatory issue. Herbicide manufacturers are looking for new strategies to expand markets, reposition products, and decrease environmental risks. Calgene's strategy to meet these needs is to genetically engineer crops to be tolerant to specific herbicides, thus expanding the range of crops in which those herbicides can be used. From the farmer's perspective, tolerant crops will offer more effective weed control at a lower cost.[26]

How will a new generation of herbicide-resistant crops affect the chemical dependency of agriculture? I raised the issue in a 1982 essay titled "Social Responsibility in an Age of Synthetic Biology"[27] and a year later in a paper titled "Biotechnology and Unnatural Selection: The Social Control of Genes."[28] This research agenda soon became a target of environmental groups. Jack Doyle warned us about the environmental consequences of programming herbicide resistance in plants in his *Altered Harvest*.[29] Periodically, this issue has been addressed by the media, as in a 1985 *Wall Street Journal* feature story titled "Quest for Pest-Resistant Crops Is Raising Ecological Concerns."[30]

The constituencies for herbicide-resistant crops, at least initially, are the biotechnology companies and the herbicide producers. Farm organizations have been either noncommittal or skeptical toward the idea. Broader constituencies, such as the mass of consumers, have had virtually no opportunity to express their opinion on the social utility of herbicide-resistant crops. It is highly unlikely, in the current regulatory climate, that the social value of these products will be debated in a national context if they are not associated with specific environmental or health hazards.

The potential adverse ecological effects of herbicide-resistant plants are of two types. First, there is a possibility that the herbicide-resistant genes could be transferred to other plants (horizontal transfer). A worst-case scenario involves the transfer of herbicide-resistant genes to weeds. This type of ecological outcome could form a basis for some regulatory agencies to prohibit dissemination of such plants. But if horizontal transfer is not expected, and no other direct adverse effects on the ecology are predicted, then from a regulatory standpoint the issue is unproblematic.

But there is a second effect that, while not related to the introduction of herbicide-resistant plants per se, is nonetheless a direct outcome. Since the introduction of these plants may result in much wider uses of herbicides, there may be an additional chemical load on the environment. For example, plant resistance to atrazines may increase the use of these persistent herbicides. Once the herbicide has been approved for use, it is difficult to establish limits on environmental loading. A study published at the University of Amsterdam distinguishes between R&D programs seeking to develop pest-resistant plants and those pursuing pesticide-resistant plants.

Rather than undertaking the longer term effort of developing pest-resistant plants, the agricultural companies may see the early opportunity of developing pesticide-resistant plants. . . . If crops—by transferring the pesticide-resistant genes into plants—could be made resistant naturally against pesticides, broad spectrum use of these new products will be guaranteed and the poison will be sprayed more freely and the quantities used will increase.[31]

Environmentalists have argued that herbicide-resistant strains are socially dysfunctional because they reinforce practices of chemical dependency in agriculture. In March 1990 a coalition of 24 representatives of public interest groups, including the National Wildlife Federation, Friends of the Earth, the Environmental Defense Fund, Pesticide Action, the National Farmer's Union, and the Rural Advancement Fund International, calling itself the Biotechnology Working Group, released a 73-page report that launched a broadside attack on the development of herbicide-tolerant plants. The study cited 27 corporations supporting research on herbicide tolerance for at least 18 crops. *Bitter Harvest* reported that herbicide-tolerant crops will result in the increased application of all types of herbicides, a spread of herbicide-resistant weeds, an increase in ground and surface water contamination, greater human exposure to toxic chemicals, increased herbicide residues in food, contamination of ecosystems, and, overall, a significant departure from sustainable agriculture. No benefits were cited to the consumer or the environment.

It is inescapable that the widespread use of herbicide-tolerant crops and trees will prolong the use of chemical herbicides for weed control. . . . From a social and economic standpoint, the introduction of herbicide-tolerant crops could exacerbate trends towards economic concentration in agriculture, the decrease in farm numbers, and the deterioration of rural communities. Applied to the Third World, such plants could have unwelcome impacts on human and environmental health and genetic diversity, as well as increasing petrochemical dependence.[32]

Despite the breadth of concerns raised by this application of biotechnology, the commercial sponsors are content and the regulatory agencies constrained by statute to keep the issues focused on the ecological effects of the new strains. Environmental activists have begun to give voice to the social utility issue in conjunction with the ecological problems, not so much as a means of influencing agency behavior, but as a way to gain broader public support for changing the ground rules under which new technologies are assessed.

There are sectors of society other than a few investor groups and agribusinesses that might benefit from herbicide-resistant crops. Industry sponsors cite two important advantages of this technology. First, herbicide-tolerant crops will support no-till agriculture and thus reduce soil erosion, a positive environmental goal. Second, herbicide manufacturers claim that the newer herbicides, for which resistance is being sought, are environmentally safer than the older, more toxic varieties.

The Biotechnology Working Group disputed these claims with several arguments: (1) current R&D is not limited to the low-dose, lower-toxicity herbicides, citing 2,4-D and atrazine as examples of two chemicals linked to cancer (see Table 11.4); (2) for weed control all combinations of herbicides are likely to be used. As long as farmers believe they can get marginal increases in productivity in the short run, multiple types of herbicide use will prevail; (3) there are no

Table 11.4
Company R&D Programs on Herbicide-Resistant Plants for Suspect Herbicides

Herbicide	Company
atrazine	Ciba-Geigy
atrazine	Calgene
2, 4–D	Diamond Shamrock
2, 4–D	Schering-Plough
Bromoxynil	Calgene
triazines	Allelix

Source: Rebecca Goldburg, Jane Rissler, Hope Shand, and Chuck Hassebrook, *Biotechnology's Bitter Harvest*. A Report of the Biotechnology Working Group, 1990.

environmentally benign herbicides. Even the so-called inert ingredients in glyphosate can be toxic to aquatic systems.

Finally, some opposition to herbicide-resistant crops is grounded on an ethical argument that does not depend upon the balance of positive to negative utility or adverse ecological factors. The German Green Party, which has affinity groups in the United States and throughout Europe, opposes herbicide resistance on the grounds that it reinforces the chemical dependency of agriculture and violates a fundamental environmental ethic, namely, the sustainable use of land. On the basis of this principle, opposition to the technology persists even in the case where substitution of lower-toxicity herbicides for their more toxic counterparts can be achieved.

The debate over herbicide-tolerant plants brings into focus the limits of technology assessment in the context of environmental law. Unlike the national dialogue over "how safe is safe" in setting environmental standards, no institutional context exists for debating secondary impacts of technology when there are no direct consequences to health and the environment. Increasing chemical dependency or the capital dependency of agriculture are not categories in the lexicon of the regulatory sector. New reform contexts have to be created to gain venue for these ideas. Table 11.5 summarizes the current status of herbicide-tolerant crops according to the seven technology-assessment categories. The social need is still uncertain, as indicated by the three possible outcomes. It is notable, however, that the nation's leading environmental groups view this product of biotechnology as a lose-lose situation for consumers and the environment.

HUMAN GROWTH HORMONE

Even without the benefit of extensive empirical study, it seems obvious that most people would prefer not to be too far below average height, everything else being equal. Extreme shortness can have a profound effect on a person's self-esteem and confidence. Thus, if there were available to adult individuals a

Table 11.5
Assessment Factors for Herbicide-Tolerant Plants

Index Variable	Index Value	Justification
EI	0	best of cases, mixed outcome.
	–	more likely, overall negative.
HE	0	best of cases mixed, if toxic herbicide use decreases.
ES	–	technological fix; increased chemical use; incompatible with goals of sustainable agriculture.
EP	+	efficiency determination by farmers and agricultural field stations; long-term consumer benefits unknown.
DJ	–	benefits skewed to highly capitalized farms.
SN	+ (win-win)	replace more toxic with less toxic herbicides; reduce cost; reduce quantity of herbicide use and soil erosion.
	0 (win-lose)	tradeoff between quantity and toxicity, soil erosion, and chemical use.
	– (lose-lose)	larger amounts of all types of herbicides; nonsustainable agriculture; more weed resistance.
MD	+	markets determined by farmer acceptance; consumer demand irrelevant.

safe and simple way to make them taller, a profitable cosmetic enterprise would flourish around the technique. However, somewhere in the mid-to-late teens growth stops and probably cannot be activated again.

But in the early years of life there are opportunities to influence the growth process, particularly through endocrine intervention. The opportunity for treating growth disorders was greatly advanced when biotechnology turned its attention to the development of human growth hormone (HGH). By transplanting the HGH gene into bacteria an indefinite number of clones can be made. Every progeny of the bacteria contains a copy of the HGH gene. When the bacteria are grown in large fermentation vats and the gene is "turned on" or expressed through the cell's regulatory biochemistry, HGH may be produced in quantities unattainable before the discovery of rDNA technology.

Pituitary dwarfism is a disease characterized by a genetic abnormality that affects the body's synthesis of growth hormone. The development of microbially produced HGH heralded the possibility of unlimited supplies of the hormone for replacement therapy in early childhood to those afflicted with the disease. Before 1958 there was no treatment for dwarfism. In that year, a procedure was developed for extracting HGH from the pituitary glands of human cadavers.[33] A two-year course of treatment required 50–100 pituitaries and cost $10–20,000. There are approximately 10–15,000 children in the United States suffering from pituitary dwarfism. Of these only about 2,000 could be supplied with HGH, which was provided free by the National Hormone and Pituitary Program. Future rates of dwarfism are likely to drop as a consequence of the increased use of prenatal screening and abortion.[34]

From an investment perspective, the market for HGH is quite limited. There are good reasons to doubt the profitability of an R&D effort to develop this product of genetic engineering if its use is restricted to dwarfs. However, when the microbial product was ready for human trials, several new uses of HGH were reported in the press, including improving the rate of healing from burns, enhancing weight loss, and slowing down the aging process. These new prospective uses of HGH have led to broader discussions about the role of the substance in clinical and nonclinical practice.

The prospect of using HGH as a replacement therapy for pituitary dwarfism is widely regarded as socially desirable, with the caveat that the hormone can be administered safely. The ethical imperative to prevent or control the outcome of this genetic affliction would almost certainly generate wide support on all social criteria. Here is a product that is needed, is desired by those afflicted with the illness, will improve distributive justice (mature dwarfs are likely to be targets of discrimination), and has a positive ethical outcome. Beyond those individuals in the population who can be helped by the replacement therapy, others should feel positive about society's role in preventing the adverse phenotypic effects of a genetic disorder.

An analogy can be drawn between society's role in providing HGH for those afflicted by pituitary dwarfism and its role in making available human insulin for those afflicted by diabetes. Both involve replacement therapeutic drugs for an inherited disorder. There are also some differences between these cases. Pituitary dwarfism is not a life-threatening disease, although it may be accompanied by an accelerated aging process. On the other hand, diabetes as a genetic disorder is life-threatening and, if not treated, is usually associated with other chronic illnesses. The differences between these cases may be relevant to balancing the risks and benefits of the therapeutic treatment, but they do not affect the ethical grounds for making treatments available to the afflicted.

HGH has also been considered for use on short-statured otherwise normal children. Under these circumstances, the child has inherited short stature. The pituitary gland may be functioning quite normally. In addition, all bodily processes are in balance and the relative growth of the anatomical elements are in

proportion. By administering growth hormone to the child it is possible to overcome some of the height limitations attributable to the inborn genetic instructions.
The issue is not hypothetical.

Human trials with growth hormone have already been carried out on a group
of 15 short but otherwise normal children. The results of the study were published
in the *New England Journal of Medicine* and stated that nearly one half of the
short, normal children grew between 2 and 4 centimeters (1–2 inches) a year
during treatment.[35]

The final paragraph of the study raises some ethical considerations about the
potential widespread use of HGH therapy.

Only long-term follow-up will establish whether sustained treatment with growth hormone
will increase the final height of these short normal children. Additional studies are needed
to determine whether some of these children have subtle abnormalities of growth hormone
or somatomedin structure, secretion or action. These preliminary results in a small sample
of very short children raise important ethical, clinical, and economic issues. Until we
have knowledge of its long-term effects and possible adverse actions in these children
and more sharply defined criteria for the selection of patients and dosage regimes, the
extrapolation of these findings to support indiscriminate treatment of short normal children
with the potent hormone is premature and unwarranted.[36]

The authors acknowledge that normal children were subjected to unknown
risks of treatment with a potent hormone (not a product of genetic engineering).
Approval for the experiment was obtained from the Committee on Human Research at UC San Francisco and the parents and patients involved. In the case
of a five year-old child, the consent of the patient is moot.

The study was cited by Tufts University biologist Ross Feldberg as an example
of the trend toward the medicalization of social problems. Feldberg questioned
the study on social and ethical grounds. These children, although they fell far
below the mean height for their ages, showed no deficiency in growth hormone
or in growth hormone receptors. Indeed, the youngest child treated was less than
five years old, hardly old enough to allow one to come to reliable conclusions
about his eventual height.

So what we have is a case of apparently healthy children being treated with
a powerful hormone in an attempt to push them toward the average height. Is
shortness per se a sickness? Shortness can indeed be a social problem. Short
individuals do suffer a variety of forms of discrimination. But is the appropriate
response to that tinkering with the individual's genes or endocrine system or
making society change?[37]

Feldberg echoed the authors' prediction that rDNA-produced human growth
hormone will radically alter the therapeutic prospects for children with short
stature because there is no theoretical limit on the availability of human growth
hormone. Since short stature is not a disease, Feldberg contends, it should not
be hostage to the disease model. Parents of short children may be more than
willing to place their diminutive offspring on a diet of HGH. We do, after all,

have cosmetic surgery of all types to modify the human anatomy toward some ideal. The goal of making short but healthy children tall constitutes a form of cosmetic endocrinology.

What are the risks of HGH therapy for short-statured normal children? In 1985, it was discovered that three people treated with cadaver-derived HGH died of Creutzfeld-Jacob Syndrome. The cause of the disease has been traced to slow viruses that live in brain tissue. It is believed that the pituitaries from which the HGH was extracted were infected by the virus.

One of the arguments for developing a genetically derived HGH is that its manufacture eliminates the threat of Creutzfeld-Jacob Syndrome. But there are other risks associated with administering a potent hormone to normal children, such as the introduction of exogenous matter from the microbial synthesis or the risks to the human system of adding more hormone to an individual's biochemical balance.

Public opinion about the social utility of treating short-statured normal children with HGH will undoubtedly be divided. Some individuals will be willing to accept some risks on behalf of their offspring to insure that he or she attains a normal height. But the risks of HGH therapy cannot be effectively determined until large numbers of children are treated with the hormone. Therefore, full disclosure of the risks to experimental subjects (in this case children) cannot be realized for years after the experiment is completed. Some people would argue that the exposure to unknown risks of this type is not warranted for cosmetic purposes. While the avoidance of short stature is a legitimate want, it is not a legitimate social need.

Others, pointing to the adverse psychological problems of excessive shortness, might view the social utility in more positive terms. For example, sociologists have determined that tallness corresponds favorably with success across many fields. The argument is advanced that there are economic and social advantages to being tall. Thus, once the therapy is available to those who want to increase their height toward the mean, it will also be available to those who wish to exceed the mean.

Since HGH may be available in the consumer market (in contrast to BGH, which is introduced into the production system) the public could have a better opportunity to express its estimate of social utility. However, the value of HGH therapy for short-statured normal children will be heavily influenced by cultural attitudes and practices toward short people. There are plentiful examples of the extremes to which people will go and the risks they will take to meet some desired measure of anatomical perfection.

The central ethical problem with the cosmetic use of HGH was enunciated by Feldberg. In cases of short stature, where clinically normal people face many forms of liability and discrimination, should they be permitted to avail themselves of endocrine therapy? "Medicalizing this 'problem' transfers the responsibility for the discrimination away from those doing the discrimination and to the victims. Are we to deal with discrimination by making short people taller or black people whiter?''[38] It might be argued that public responsibility ends when individuals are adequately informed about the risks of cosmetic endocrinology. However, this area of social ethics has not been resolved. Which decisions are legitimately the responsibility of the government and which, for better or worse, are left to the individual?

When growth hormone therapy is used to combat pituitary dwarfism, a clinical norm is introduced based upon the body's production of the hormone. This norm, for the most part, is independent of social and cultural factors. However, when HGH is used for cosmetic purposes, there is no stable norm. Clinical and social norms are mixed. If height is associated with social or athletic advantage, then the advantage can only be exploited if one individual has access to the therapy and others do not.

The ethical dimensions of the argument about clinical and cosmetic uses of HGH may be illuminated further by the work of the philosopher Immanuel Kant. In evaluating whether a type of behavior (framed as a maxim) is morally justifiable, Kant applied a universalization criterion. If the maxim can be universalized without contradiction, then, Kant concluded, it can serve as the basis of action for all individuals facing similar situations. As an example, lying to gain an advantage cannot be universalized without undermining the aims of the perpetrator of the lie.

Consider the following maxims: (1) all pituitary dwarfs may be treated with HGH to insure that a normal amount of hormone is available for bone development; (2) children who wish to attain a normal height or a height advantage may be treated with HGH. The first maxim can be universalized without any contradictory results. However, suppose we imagine that everyone carried out the second maxim and all small children were provided with HGH. Assuming that the therapy worked for normal children, the advantage for any individual would be negated. The norm of acceptable height or mean height shifts when we universalize the maxim; therefore, any relative advantage is lost.

The assessment heuristic applied to HGH illustrates the factors involved in a full-scale public debate. Even at this stage, there are significant differences in how people view the social utility of HGH treatment for normal children. There is anticipation of sizable HGH markets despite the vast ignorance that exists about the long-term health consequences. In addition, the ethical problems of using drugs to redress the perceived maldistribution of human genes poses another significant problem.

The following scenario describing an HGH future was reported in *Science*.

In the near future, say investigators who study human growth, it is almost certain that many affluent parents of short children will have their children treated with human growth hormone as a matter of course. The children then might grow up to be "appropriately" tall. And there is also a likelihood that obese people will take growth hormone when they diet so that they will lose fat and not muscle tissue. Growth hormone may even be used to retard aging—it may prevent wrinkles and the sort of fat distribution that occurs in old people. Athletes are already taking the drug, obtaining it on the black market, because there is evidence that it may help build muscle tissue.[39]

HGH is beginning to sound like a miracle product, similar to the way interferon was publicized several years earlier. The excitement over HGH was greatly reinforced during the summer of 1990 when a widely publicized study in the *New England Journal of Medicine* reported a reversal in the effects of aging of healthy elderly men treated with HGH. The subtext in the *Boston Globe* read "Researchers say it [HGH] turned clock back 10 to 20 years for a group of men 61–81."[40] Some physicians responded to the report with stern warning against cosmetic endocrinology on the basis of what is currently known, somewhat like the warnings against the use of steroids for improving muscle tone.

The prospect of large-scale marketing of HGH for short normal children or normal aging adults is reminiscent of the widespread application to women of the synthetic estrogen diethylstilbestrol (DES) as a postcoital contraceptive. Only after years of use was it discovered that DES daughters had an increased risk of vaginal cancer.[41] Eventually, the use of the hormone in humans and animals was significantly curtailed. Class action suits were mounted by those adversely affected in an attempt to draw compensation from the DES manufacturers. It may be impossible to fully evaluate the long-term effects of HGH on a healthy population without exposing the population to an inestimable risk.

FROM TECHNOLOGY ASSESSMENT TO
SOCIAL GUIDANCE

Biotechnology has been responsible for a myriad of technological innovations covering multiple sectors of the economy. These innovations have been amply summarized in this and other works. At the root of these innovations is the conscious rearrangement of biological forms (biotechnics) through genetic controls (gentechnics). Microchanges in the fundamental chemical units of living entities are reflected in the macrochanges taking place in the reconfiguration of the industrial sector. The new symbols applied to genetic science speak to a mechanistic and instrumentalist vision of living things. Yanchinski's terminology "setting genes to work" and Yoxen's "life as a productive force" are expressive of the links between the science of living forms and the technology of manufacture that have become the signature of the biotechnological revolution.[42] Goodman et al. use the term "bio-industrialization" to describe the "increasing transfer and interchangeability of both industrial processes and inputs between the food, chemicals, and pharmaceuticals sectors."[43]

Innovation, investment, and development in applied genetics have been robust. The fervor of bioindustrialization is as strong in private as in public sector institutions. It can be felt at the state, federal, and international levels. Not since the discovery of antibiotics has there been this level of expectation associated with biomedical developments. Not since the introduction of hybrid seeds has there been as much excitement within industrial agriculture.

The aggressive exploitation of genetic science for practical ends is by and large a healthy development. But equally important are the processes and social mechanisms through which selection of potential applications is carried out. I have argued that the current methods of assessing the impacts of biotechnology and for choosing among alternative technological paths have not been commensurate with the incentives to develop and market new products and to transform methods of production. There are several reasons for this.

First, there is a confusion of roles. Technological innovation of commercial products should reside primarily with the private sector. The public sector roles should serve to protect society from misdirected technologies. Currently, public sector institutions are too closely identified with the development side of biotechnology. This has resulted in conflicts within federal and state governments over the appropriate regulatory stance.

Second, universities have lost their role as independent sources of analysis, valuation, and assessment of new biotechnologies. The academic research community in applied genetics has become integrated into a system of commercial development that has brought industry, government, and the university into an unprecedented peacetime partnership.

Third, the biotechnology revolution has emerged at a time when the social demands on technology are far more complicated than they once were. The social guidance systems have not kept pace with social attitudes. Productivity is only one of several competing values that form part of the public's assessment agenda for technological change. Greater attention is being placed on secondary impacts of technology beyond its direct effects on human health. A new powerful metaphor, Gaia, the organism of earth, is placing new demands on innovations in manufacture and production.

There is also a new global economic perspective on the effects of technological change. If we modify our packaging materials or develop a microbial process for making cocoa, we may inadvertently but predictably accelerate the rapid depletion of the world's rain forests. These considerations, once the province of fringe ecotopians, have become normalized into public values. Thus, our assessment methods for technology are deficient because social expectations have changed. Periodically, there are examples where the regulatory sector is baffled by a public outcry over what is viewed as an orderly and statutorily correct response to a problem. For example, ALAR, a chemical used to control the ripening time of apples and shown to cause cancer in animals, was eliminated from use when significant segments of the public refused to purchase produce sprayed with the chemical. A similar reaction prompted emergency restrictions on the use of the pesticide ethylene dibromide (EDB) in grain products.

I have shown that some of the concerns expressed about products derived by genetic engineering techniques fall outside the responsibility of regulatory bodies. Where a product has questionable or potentially negative human health effects or is a clear and present ecological hazard, it has issue-legitimacy within the existing regulatory sectors. However, for those products or technologies with second-order environmental effects, redistributive effects, or that raise ethical dilemmas, there are no natural places toward which public debate is channeled.

Our federal structure is not currently designed for the public to direct the course of technology, for constituencies to question the social utility of products that are not otherwise deemed hazardous, to evaluate the ecological impacts of innovations in production, to propose directions for technological development, or to solve complex ethical problems associated with new technologies. A market-dominated innovation system makes it extremely difficult for socially guided R&D programs to evolve. There is little guarantee, thus far, that the potential biotechnology offers will correlate with the hierarchy of social needs. Our examples are selective and do not tell the whole story. There are many applications of biotechnology that are not problematic and contribute quality or efficiency to systems of manufacture or the treatment of disease. Those are not the outcomes of biotechnology that place our current system of technology assessment to the test. The cases chosen in this analysis illustrate the complex problems of technological choice that biotechnology puts before us.

Too many questions related to the effects of biotechnology are defined outside the responsibility of government. Too many of our agencies of government conceive of their role as promoting innovation and development rather than assessment and selectivity. Too many of those in whom we expect objectivity have vested interests in the financial success of a technology. The inevitable outcome of this situation is that organized efforts by nongovernmental groups give up working with federal agencies and work directly with the public, and scientists lose their special status in society. We need new institutional models to examine the total system impact of innovations in biotechnology in a manner that responds to multiple constituencies. The assessment of innovations in biotechnology must rise above the current fragmentary approach defined by the regulatory sphere. Comments I made nearly a decade ago are as relevant today. "The developments in a field bursting with innovative ideas and [unexplored] potential will put to the test the social guidance systems we presently have. But more so, they will test the moral and scientific wisdom of technologically advanced countries on their capacity to counteract the adverse effects of genetic technology before they are realized and become part of the social and economic infrastructure of society."[44]

Notes

PREFACE

1. David F. Noble, *Forces of Production* (New York: Alfred A. Knopf, 1984), xi.
2. John R. Fowle, III, ed., *Application of Biotechnology: Environment and Policy Issues* (Boulder, CO: Westview Press, 1987).

CHAPTER 1

1. Francis Bacon, *The New Atlantis*, ed. H. Goitein (London: George Routledge & Sons, n.d.), 242.
2. Edward Yoxen, *The Gene Business* (New York: Harper & Row, 1984), 4.
3. Bernard Cohen, *Revolution in Science* (Cambridge, MA: Belknap Press of Harvard University Press, 1985).
4. Lewis Mumford, *Technics and Civilization* (New York: Harcourt, Brace & Co., 1934).
5. Ibid., 217.
6. Ibid., 109–10.
7. Ibid., 243.
8. Ibid., 353.
9. Ibid., 254.
10. Yoxen, 15.
11. Richard Dawkins, *The Selfish Gene* (New York: Oxford University Press, 1976).
12. The symbol of the cell as machine has been captured by popular science writers. See, for example, John Elkington, *The Gene Factory* (New York: Carroll & Graf, 1985).
13. Rick Gore, "The Awesome Worlds Within a Cell," *National Geographic* 150 (September 1976): 355–95.
14. David A. Hopwood, "The Genetic Programming of Industrial Microorganisms," *Scientific American* 245 (September 1981): 91–102.
15. Peter Singer, *Animal Liberation* (New York: New York Review, 1975), 98.
16. Yoxen, 18.
17. Yoxen, 19.

18. Dawkins.

19. Lewis Mumford, *The Culture of Cities* (New York: Harcourt, Brace & Co., 1938), 496.

20. Yoxen; J. Kloppenburg, Jr., and M. Kenney, "Biotechnology, Seeds, and the Restructuring of Agriculture," *Insurgent Sociologist* 12 (Summer 1984): 3–18; Jack Doyle, *Altered Harvest* (New York: Viking, 1985).

21. Office of Technology Assessment, *Technology, Public Policy, and the Changing Structure of American Agriculture* (Washington, DC: U.S. Government Printing Office, 1986), 112.

22. Doyle, 294.

23. See Ruth Hubbard, *The Politics of Women's Biology* (New Brunswick, NJ: Rutgers University Press, 1990), 81. "Genes do not reproduce and DNA does not replicate itself, as they are sometimes said to do. Their reproduction or replication happens as part of the metabolism of living cells."

24. Richard Levins and Richard Lewontin, *The Dialectical Biologist* (Cambridge, MA: Harvard University Press, 1985).

25. Martin Sherwin writes: "Over a period of less than three years the [atomic bomb] program cost over $2-billion, required the construction and use of thirty-seven installations in nineteen states and Canada, employed approximately 120,000 persons, and absorbed a large proportion of the nation's scientific and engineering talent." *A World Destroyed* (New York: Alfred A. Knopf, 1975), 42.

26. Charles Piller and Keith R. Yamamoto, *Gene Wars* (New York: Beech Tree Books, 1988).

27. Langdon Winner, *Autonomous Technology: Technics Out of Control as a Theme in Political Thought* (Cambridge, MA: MIT Press, 1977).

CHAPTER 2

1. Daniel E. Koshland, Jr., "Excursions in Biotechnology," *Science* 229 (20 September 1985): 1191.

2. Barry Commoner, "Bringing Up Biotechnology," *Science for the People* 19 (March/April 1987): 9.

3. Office of Technology Assessment, *Commercial Biotechnology: An International Analysis* (Washington, D.C.: U.S. Government Printing Office, 1984), 3.

4. David Webber, "Biotechnology Moves into the Marketplace," *Chemical & Engineering News* 62 (16 April 1984): 12. "There is no such thing as a biotechnology industry."

5. OTA, *Commerical Biotechnology*, 72.

6. OTA, *New Developments in Biotechnology: U.S. Investment in Biotechnology— Special Report 4*, OTA–BA–360 (Washington, DC: U.S. Government Printing Office, 1988), 78.

7. Arnold L. Demain and Nadine A. Solomon, "Industrial Microbiology," *Scientific American* 245 (September 1981): 66–75.

8. Carl S. Pederson, *Microbiology of Food Fermentations* (Westport, CT: Avi Publishing Co., 1971), 260.

9. Demain and Solomon, 67.

10. Ibid.

11. Pederson, 66.

12. Ibid., 260.

13. C. W. Hesseltine, "A Millenium of Fungi, Food, and Fermentation," *Mycologia* 57 (March/April 1965): 149–97.

14. Ibid., 175.

15. Demain and Solomon, 67.

16. Ibid., 74.

17. K. Sakaguchi, T. Uemura, and S. Kinoshita, eds. *Biochemical and Industrial Aspects of Fermentation* (Tokyo: Kodansha, 1971).

18. Edward Yoxen, *The Gene Business* (New York: Harper & Row, 1984), 16.

19. Marc Lappé, *Broken Code: The Exploitation of DNA* (San Francisco: Sierra Club Books, 1984), xi.

20. OTA, *Commercial Biotechnology*, 65.

21. OTA, *New Developments*, 2.

22. Spyros Andreopoulos, "Sounding Board: Gene Cloning by Press Conference," *The New England Journal of Medicine* 302 (27 March 1980): 743–45.

23. Harold M. Schmeck, Jr., "Natural Virus-Fighting Substance is Reported Made by Gene-Splicing," *New York Times*, 17 January 1980.

24. Gene Bylinksy, "DNA Can Build Companies Too," *Fortune* 101 (16 June 1980): 144–54.

25. Mark Czarneck, "The Selling of DNA," *World Press Review* (September 1980), 34.

26. A. Gerassimos et al., "Recombinant Alfa–2b-Interferon Therapy in Untreated, Stages A and B Chronic Lymphocytic Leukemia," *Cancer* 61 (1988): 869–73; David Goldstein et al., "Interferon Therapy in Cancer," in *Principles of Cancer Biotherapy*, ed. Robert K. Oldham (New York: Raven Press, 1987), 247–72.

27. Frederick G. Hayden et al., "Intranasal Recombinant Alfa–2b Interferon Treatment of Naturally Occurring Common Cold," *Antimicrobial Agents and Chemotherapy* 32 (February 1988): 224–31.

28. Robert Teitelman, *Gene Dreams* (New York: Basic Books, 1989), 211.

29. OTA, *Commercial Biotechnology*, 121.

30. Ibid., 185.

31. "Only the Beginning of Startling Changes," *Business Week* (6 July 1981), 56.

32. Webber, "Biotechnology Moves into the Marketplace."

33. OTA, *Commercial Biotechnology*, 270.

34. OTA, *New Developments*, 81.

35. Ibid., 10.

36. Zsolt Harsanyi, *Investment Strategies in Biotechnology: The Race to Commercialization* (New York: Eic/Intelligence, 1984), 10.

37. OTA, *New Developments*, 5.

38. Ibid., 9.

39. OTA, *Commercial Biotechnology*, 278.

40. Martin Kenney, *Biotechnology: The University-Industrial Complex* (New Haven: Yale University Press, 1986), 166.

41. Joseph W. Bartlett and James V. Siena, "Research and Development Limited Partnerships as a Device to Exploit University-Owned Technology," *Journal of College and University Law* 10 (Spring 1983–84): 435–54.

42. OTA, *Commercial Biotechnology*, 96.

43. Ibid., 96.

44. OTA, *New Developments*, 68.

45. Ibid., 61.

46. Wisconsin Department of Natural Resources, *State Agency Biotechnology Report and Survey Results: Legislative and Regulatory Activities*, Final Report, 1987.

47. OTA, *New Developments*, 71.

48. Kenney.

49. OTA, *Commercial Biotechnology*; OTA, *New Developments*.

50. OTA, *New Developments*, 5.

51. OTA, *Commercial Biotechnology*, 92–93.

52. OTA, *New Developments*, 83.

53. OTA, *Commercial Biotechnology*, 91.

54. Charles Piller and Keith R. Yamamoto, *Gene Wars* (New York: Beech Tree Books, 1988).

55. Veronica Fowler, "Ethics Debate Encircles Biotech Research," *Des Moines Register*, 27 July 1987.

CHAPTER 3

1. Quoted in William Booth, "Animals of Invention," *Science* 240 (6 May 1988): 718.

2. Tom Regan, "Moratorium on Higher Life Forms" [letter to the editor], *New York Times*, 24 April 1988.

3. U.S. Constitution, Article 1, Section 8.

4. J. Kloppenburg, *The First Seed: The Political Economy of Plant Biotechnology, 1492–2000* (New York: Cambridge University Press, 1987).

5. Office of Technology Assessment, *New Developments in Biotechnology: Ownership of Human Tissues and Cells—Special Report 1*, OTA–BA–337 (Washington, DC: U.S. Government Printing Office, 1987).

6. House Committee on the Judiciary, Subcommittee on Courts, Civil Liberties, and the Administration of Justice, *Patents and the Constitution: Transgenic Animals*, hearings, 11 June, 22 July, 21 Aug., and 5 Nov. 1987 (Washington, DC: U.S. Government Printing Office, 1988).

7. House Committee on the Judiciary, report on H.R. 4970, *Transgenic Animal Patent Reform Act*, 26 August 1988.

8. Office of Technology Assessment, *New Developments in Biotechnology: Patenting Life—Special Report 5*, OTA–BA–370 (Washington, DC: U.S. Government Printing Office, 1989), 87.

9. Ibid., 117.

10. M. W. Fox, "Effect of Growth Hormone on Cows" [letter], *Science* 233 (19 September 1986): 1247.

11. M. W. Fox, "Genetic Engineering: Human and Environmental Concerns" (briefing paper, 1983).

12. Veronica Fowler, "Ethics Debate Encircles Biotech Research," *Des Moines Register*, 27 July 1987.

13. House Committee on the Judiciary, *Transgenic Animal Patent Reform Act*, 36.

14. B. E. Rollin, " 'The Frankenstein Thing': The Moral Impact of Genetic Engineering of Agricultural Animals on Society and Future Science," in *Genetic Engineering of Animals*, ed. A. Hollaendor and C. Wilson (New York: Plenum, 1986), 285–97.

15. Baruch A. Brody, "Animal Patents: The Legal, Economic and Social Issues" (Paper presented at Cornell University, 5–6 December 1988).

CHAPTER 4

1. Arnold S. Relman, "The Academic Corporate Merger in Medicine: A Two-Edged Sword," in *Genetics and the Law III*, ed. Aubrey Milunsky and George J. Annas (New York: Plenum, 1985), 39–43.

2. Sheldon Krimsky and David Baltimore, "The Ties that Bind or Benefit," *Nature*, 283 (10 January 1980): 130.

3. Arthur H. Westing, ed., *Environmental Warfare* (London: Taylor and Francis, 1984), idem, *Herbicides in War* (London: Taylor and Francis, 1984).

4. Loren Baritz, *The Servants of Power: A History of the Use of Social Science in American Industry* (Middletown, CT: Wesleyan University Press, 1960).

5. Steven Halliwell, "Columbia: An Explanation," in *The New Left*, ed. Priscilla Long (Boston: Porter Sargent, 1969), 207.

6. Noam Chomsky, "Knowledge and Power: Intellectuals and the Welfare-Warfare State," in *The New Left*, ed. Priscilla Long, 172–99.

7. David Horowitz, "Billion Dollar Brains," *Ramparts* (May 1969): 36–44.

8. Charles Schwartz, "The Corporate Connection," *Bulletin of the Atomic Scientists* 31 (October 1975): 15.

9. Ibid.

10. Sheldon Krimsky, "The Transformation of the American University," *Alternatives* 14 (May/June 1987): 20–29. See also Sheldon Krimsky, "University Entrepreneurship and Public Purpose," in *Biotechnology: Professional Issues and Social Concerns*, ed. P. DeForest et al. (Washington, DC: American Association for the Advancement of Science, 1988), 34–42; Philip L. Bereano, "Making Knowledge a Commodity: Increased Corporate Influence on Universities," *IEEE Technology and Society Magazine* 5 (December 1986): 8–17.

11. National Science Foundation, *Federal Funds for Research and Development*, various years, including 1986, 1987, 1988.

12. Office of Technology Assessment, *New Developments in Biotechnology: U.S. Investment in Biotechnology—Special Report 4*, OTA–BA–360 (Washington, DC: U.S. Government Printing Office, 1988).

13. Ibid., 41.

14. Charles Piller and Keith R. Yamamoto, *Gene Wars* (New York: Beech Tree Books, 1988), 155. Susan Wright, "The Military and the New Biology," *Bulletin of the Atomic Scientists* 41 (May 1985): 10–16; idem, "Chemical/Biological Weapons: The Buildup That Was," *Bulletin of the Atomic Scientists* 45 (January/February 1989): 52–56. See also Susan Wright, ed., *Preventing a Biological Arms Race* (Cambridge, MA: MIT Press, 1990).

15. G. A. Keyworth, II, "Federal R&D and Industrial Policy," *Science*, 220 (10 June 1983): 1122–25.

16. Judith A. Johnson, "Biotechnology: Commercialization of Academic Research," Issue Brief No. IB81160 (Washington, DC: Congressional Research Service, Library of Congress, 18 August 1982).

17. Norton D. Zinder and Jackie Winn, "A Partial Summary of University-Industry Relationships in the United States," *Recombinant DNA Technical Bulletin* 7 (March

1984): 8–19. Colin Norman, "Electronic Firms Plug into Universities," *Science* 217 (6 August 1982): 511–14.

18. Krimsky and Baltimore, 130–31.

19. Ronald Rosenberg, "Collaborative in First Public Offering with 1.5m Shares," *Boston Globe*, 10 December 1981.

20. Katherine Bouton, "Academic Research and Big Business: A Delicate Balance," *New York Times Magazine*, 11 September 1983. Will Lepkowski, "University/Industry Research Ties Still Viewed with Concern," *Chemical & Engineering News* 62 (25 June 1984): 7–11. Sally Macdonald and Lee Moriwaki, "Corporate Gifts: Is the Camel Getting His Nose in the Tent," *Seattle Times/Seattle Post-Intelligencer*, 18 November 1984. Ann Crittenden, "Industry's Role in Academia," *New York Times*, 22 July 1981. Leon Lindsay, "Universities + Business =," *Christian Science Monitor*, 29 October 1982. Barbara J. Culliton, "Academe and Industry Debate Partnership," *Science* 219 (14 January 1983): 150–51. David M. Kiefer, "Forging New and Stronger Links between University and Industrial Scientists," *Chemical & Engineering News* 58 (8 December 1980): 38–51. David F. Noble and Nancy E. Pfund, "Business Goes Back to College," *Nation* 233 (20 September 1980): 246–52. Anne C. Roark, "Academic Ties Face Challenge," *Harvard Magazine* (January/February 1981): 15–19. Stephanie Yanchinski, "Universities Take to the Marketplace," *New Scientist*, (3 December 1981): 675–77. Melinda Walsh, "University, Industry Ties Create a Touchy Situation," *Daily Democrat*, Woodland-Davis, California, 1 July 1982. Matthew Oyster, "Biotechnology at Berkeley: Up for Grabs?" *California Monthly* (June-July 1982). Virginia Inman, "Professors are Taking More Consulting, with College Approval," *Wall Street Journal*, 31 March 1983.

21. David E. Sanger, "Corporate Links Worry Scholars," *New York Times*, 17 October 1982. M. R. Montgomery, "The Selling of Science," *Boston Globe Magazine*, 21 August 1983. William Boly, "The Gene Merchants," *California Magazine*, September 1982. Richard A. Knox, "Biotech: Secrecy and Questions in Gold-Rush Atmosphere," *Boston Globe*, 15 September 1981. Stephanie Yanchinski, "Deafening Silence from Genetic Engineers about Commercial Threats," *New Scientist* (1 October 1981): 4. Glaucon, "Biology Loses Her Virginity," *New Scientist* (18 December 1980): 826. Joyce Christopher, "New Companies Could Turn Academics into Tycoons," *New Scientist* (28 May 1981): 542. Barbara Culliton, "Biomedical Research Enters the Marketplace," *New England Journal of Medicine* 20 (14 May 1981): 1195–1201. Jeffrey L. Fox, "Can Academia Adapt to Biotechnology's Lure?" *Chemical & Engineering News* 59 (12 October 1981): 39–44.

22. House Committee on Science and Technology, Subcommittee on Investigations and Oversight, and Subcommittee on Science, Research and Technology, *Commercialization of Academic Biomedical Research*, hearings, 8–9 June 1981 (Washington, DC: U.S. Government Printing Office, 1981); idem, *University/Industry Cooperation in Biotechnology*, hearings, 16–17 June 1982 (Washington, DC: U.S. Government Printing Office, 1982). House Committee on Energy and Commerce, Subcommittee on Oversight and Investigations, *Biotechnology Development*, hearings, 18 December 1985 (Washington, DC: U.S. Government Printing Office, 1986).

23. They drew an initial sample of 1,997 faculty in a two-step process. First they selected 40 universities from the top 50 recipients of federal grants in the United States. Second, they chose 3,180 life sciences faculty members from the 40 institutions by culling catalogs. From that list 1,594 were randomly selected. They also selected a comparison

group of 403 nonlife scientists. After screening for eligibility, they ended up with 993 respondents from the life sciences and 245 in other areas.

24. Alexander G. Bearn, "The Pharmaceutical Industry and Academe: Partners in Progress," *American Journal of Medicine* 71 (July 1981): 81–88.

25. Keith R. Yamamoto, "Faculty Members as Corporate Officers: Does Cost Outweigh Benefits?" in *From Genetic Experimentation To Biotechnology—The Critical Transition*, ed. W. J. Whelan and Sandra Black (Chichester, England: Wiley, 1982), 198.

26. David Blumenthal et al., "Industrial Support of University Research in Biotechnology," *Science* 231 (17 January 1986): 242–46.

27. David Blumenthal et al., "University-Industry Research Relationships in Biotechnology: Implications for the University," *Science* 232 (13 June 1986): 1361–66.

28. Ibid., 1364.

29. This observation was made by Bernard Davis, Harvard Medical School. See A. Milunsky and G. J. Annas, eds., *Genetics and the Law* (New York: Plenum, 1985), 67. Science journalist Philip Hilts wrote: "It is already apparently true that there is no notable biologist in this field anywhere in America who is not working in some way for business. I interviewed some two dozen of the best molecular biologists in the country and found none." *Scientific Temperaments* (New York: Simon and Schuster, 1982), 185.

30. The discussion of university–industry linkages in this chapter was adapted from Sheldon Krimsky, James G. Ennis, Robert Weissman, "Academic–Corporate Ties in Biotechnology: A Quantitative Study," *Science, Technology & Human Values* 16 (July 1991).

31. Philip L. Bereano, "Making Knowledge a Commodity: Increased Corporate Influence of Universities," *IEEE Technology and Society Magazine* 5 (December 1986):8–17.

CHAPTER 5

1. Sharon McAuliffe and Kathleen McAuliffe, *Life for Sale* (New York: Coward, McCann & Geoghegan, 1981), 114.

2. Gordon Bultena and Paul Lasley, "The Dark Side of Agricultural Biotechnology: Farmers' Appraisals of the Benefits and Costs of Technological Innovation," in *Agricultural Bioethics*, ed. S. M. Gendel et al. (Ames: Iowa State University Press, 1990), 107–8.

3. Earle H. Harbison, Jr., "Biotechnology Commercialization—An Economic Engine for Future Agricultural and Industrial Growth" (Presentation at the World Economic Forum, Davos, Switzerland, 5 (February 1990).

4. Ibid.

5. Francis Bacon, *The New Atlantis*, ed. H. Goitein (London: George Routledge & Sons, n.d.).

6. Ibid., 243.

7. Ibid.

8. Philip J. Pauly, *Controlling Life: Jacques Loeb and the Engineering Ideal in Biology* (New York: Oxford University Press, 1987), 94.

9. Ibid., 108.

10. Quoted in ibid., 50–51.

11. Steve Olson, *Biotechnology: An Industry Comes of Age* (Washington, D.C.: National Academy Press, 1986), 39–42. A transgenic mouse was pictured on the front cover

of *Science* 222 (18 November 1983). This was the first example of a human gene expressed in another animal.

12. Carolyn Merchant, *The Death of Nature* (New York: Harper and Row, 1980).

13. Sheldon Krimsky, "Patents for Life Forms *Sui Generis*: Some New Questions for Science, Law, and Society," *Recombinant DNA Technical Bulletin* 4 (April 1981): 11–15.

14. A. M. Chakrabarty, "Microbial Genetic Engineering and Environmental Pollution," in *Genetic Engineering* vol. 4, proceedings of an international conference, 6–10 April 1981 (Seattle, WA: Battelle Memorial Institute, 1981). Leslie Roberts, "Discovering Microbes with a Taste for PCBs," *Science* 237 (28 August 1987): 975–77.

15. Charles W. Stevens, "A Germ of an Idea May Give Microbes Taste for Dangers," *Wall Street Journal*, 14 March 1989, B4.

16. Another California firm, Advanced Genetic Sciences, promoted its product, Frostban, with a similar environmental appeal. "Frostban is a natural approach to frost protection that employs nature's own methods. It is an alternative to synthetic chemicals which are indiscriminate and ecologically destructive." See Sheldon Krimsky and Alonzo Plough, *Environmental Hazards*: *Communicating Risks as a Social Process* (Dover, MA: Auburn House, 1988), 104.

17. Joseph Haggin, "Monsanto Uses Genetic Engineering to Solve Agricultural Problems," *Chemical & Engineering News* 66 (15 February 1988): 28–33.

18. Douglas McCormick, "How Biotech is Doing with Its Nitrogen Fixation," *Bio/ Technology* 6 (April 1988): 383–85.

19. Graeme Hobson, "How the Tomato Lost Its Taste," *New Scientist*, (29 September 1988): 46–49. Quoted from the abstract in A. I. Bowker, *Telegen Reporter Annual*, vol. 7, 1988, p. 314.

20. Kenneth A. Barton and Winston J. Brill, "Prospects in Plant Genetic Engineering," *Science* 219 (11 February 1983): 671–76.

21. Ibid., 672.

22. The impact of biotechnology on the Third World was the subject of a special report of the Belgium-based International Coalition for Development Action. Henk Hobbelink, *New Hope or False Promise? Biotechnology and Third World Agriculture* (Brussels: International Coalition for Development Action, April 1987).

23. Guido Ruivenkamp, "Biotechnology: The Production of New Relations within the Agroindustrial Chain of Production" (Paper delivered to the World Food Assembly, Rome, 12–15 November 1984), 17.

24. Steve Olson, *Biotechnology: An Industry Comes of Age* (Washington, DC: National Academy of Sciences, 1986).

25. Sheldon Krimsky, "Risky Science: Is Anybody Watching the Experimental AIDS Mouse?" *Scientist* 2 (16 May 1988): 12–13.

26. Margaret Mellon, *Biotechnology and the Environment* (Washington, DC: National Wildlife Federation, 1988).

27. Susan R. Jones, "Can Bugs Pump Life into Old Oil Fields," *Chemical Week* 140 (18 March 1987): 59–60.

28. Sheldon Krimsky and A. H. Fraenkel, "United States and Canadian Governmental Regulations Concerning Biohazardous Effluents," in *Comprehensive Biotechnology*, vol. 4, *The Practice of Biotechnology*, ed. Murray Moo-Young (New York: Pergamon, 1986), 609–31. Robert H. Zaugg and J. R. Swarz, "Industrial Use of Applied Genetics and Biotechnologies," *Recombinant DNA Technical Bulletin* 5 (March 1982): 7–13.

CHAPTER 6

1. National Academy of Sciences, Committee on the Introduction of Genetically Engineered Organisms into the Environment, *Introduction of Recombinant DNA-Engineered Organisms into the Environment: Key Issues* (Washington, DC: National Academy Press, 1987), 6–7. Members of the committee involved in the preparation of this consensus report were: Arthur Kelman (Chair), University of Wisconsin-Madison; Wyatt Anderson, University of Georgia; Stanley Falkow, Stanford University; Nina Federoff, Carnegie Institution of Washington; Simon Levin, Cornell University.

2. Martin Alexander, "Ecological Consequences: Reducing the Uncertainties," *Issues in Technology and Society* 1 (Spring 1985): 57–68.

3. R. E. Kasperson et al., "The Social Amplification of Risk: A Conceptual Framework," *Risk Analysis* 8 (1988): 177–87.

4. Winston Brill, "Genetic Engineering in Agriculture" [letter], *Science* 229 (12 July 1985): 115–17.

5. Martin Kenney, *Biotechnology: The University-Industrial Complex* (New Haven: Yale University Press, 1986).

6. The inventory of biotechnology firms I collected for this work exceeds that compiled by the Office of Technology Assessment in its 1984 study. See Office of Technology Assessment, *Commercial Biotechnology: An International Analysis*, OTA–BA–218 (Washington, DC: U.S. Government Printing Office, 1984), 93.

7. Geoffrey M. Karny, "Regulation of Genetic Engineering: Less Concern about Frankensteins but Time for Action on Commercial Production," *University of Toledo Law Review* 12 (Summer 1981): 815–68.

8. John Tooze, "International and European Regulation of Recombinant DNA Research," *University of Toledo Law Review* 12 (Summer 1981): 869–90.

9. Sheldon Krimsky, *Genetic Alchemy: The Social History of the Recombinant DNA Controversy* (Cambridge, MA: MIT Press, 1982), 312–37.

10. Office of Technology Assessment, *Impacts of Applied Genetics: Microorganisms, Plants and Animals*, OTA–HR–132 (Washington, DC: U.S. Government Printing Office, 1981), 219.

11. Ibid., 218.

12. National Institutes of Health, "Guidelines for Research Involving Recombinant DNA Molecules," *Federal Register* 45 (January 1980): 6724–49.

13. National Institutes of Health, "Guidelines for Research Involving Recombinant DNA Molecules," *Federal Register* 43 (December 1978): 60108–31.

14. National Institutes of Health, "Physical Containment for Large-Scale Uses of Viable Organisms Containing Recombinant DNA Molecules," *Federal Register* 45 (April 1980): 24968–71.

15. Recombinant DNA Advisory Committee, minutes of meeting, 25–26 September 1980.

16. Shawna Vogel, "Biogenetic Waste: A Federal and Local Problem," *GeneWatch* 2 (1985): 7–9.

17. National Institutes of Health, "Guidelines for Research Involving Recombinant DNA Molecules," *Federal Register* 48 (June 1983): 24556–81.

18. E. Milewski and B. Talbot, "Proposals Involving Field Testing of Recombinant DNA Containing Organisms," *Recombinant DNA Technical Bulletin* 6 (December 1983): 141–45.

19. Sheldon Krimsky, "Research under Community Standards: Three Case Studies," *Science, Technology & Human Values* 11 (Summer 1986): 14–33.

20. Jeremy Rifkin, *Algeny* (New York: Viking, 1983); idem, *Declaration of a Heretic* (Boston: Routledge & Kegan Paul, 1985), 74.

21. National Academy of Sciences, *Research with Recombinant DNA* (Washington, DC: National Academy of Sciences, 1977), 18–21.

22. House Committee on Science and Technology, Subcommittee on Investigations and Oversight, *The Environmental Implications of Genetic Engineering* (Washington, DC: U.S. Government Printing Office, 1984).

23. F. E. Sharples, "Spread of Organisms with Novel Genotypes: Thoughts from an Ecological Perspective," *Recombinant DNA Technical Bulletin* 6 (June 1983): 43–56. James Gillett, ed., *Prospects for Physical and Biological Containment of Genetically Engineered Organisms*, proceedings of the Shackelton Point Workshop on Biotechnology Impact Assessment, 1–4 October 1985 (Ithaca, NY: Ecosystems Research Center, Cornell University, 1987).

24. Robert K. Colwell et al., "Genetic Engineering in Agriculture," *Science* 229 (12 July 1985): 111–12.

25. Eugene P. Odum, "Biotechnology and the Biosphere," *Science* 229 (12 July 1985): 112–13.

26. Harlyn O. Halvorson, David Pramer, and Marvin Rogul, eds., *Engineered Organisms in the Environment: Scientific Issues*, proceedings of a cross-disciplinary symposium held in Philadelphia, 10–13 June 1985 (Washington, DC: American Society for Microbiology, 1985). Frances E. Sharples, "Regulation of Products from Biotechnology," *Science* 235 (13 March 1987): 1329–32. Bernard D. Davis, "Bacterial Domestication: Underlying Assumptions," *Science* 35 (13 March 1987): 1329, 1332–35.

CHAPTER 7

1. Kurt Vonnegut, Jr., *Cat's Cradle* (New York: Dell, 1963), 160.

2. Mary Shelley, *Frankenstein* (New York: Bantam, 1967); Michael Crichton, *The Andromeda Strain* (New York: Alfred E. Knopf, 1969).

3. L. R. Maki, et al. "Ice Nucleation Induced by *Pseudomonas syringae*," *Applied Microbiology* 28 (1974): 456.

4. S. E. Lindow, "Ecology of *Pseudomonas syringae* Relevant to the Field Use of Ice-Deletion Mutants Constructed in Vitro for Plant Frost Control," in *Engineered Organisms in the Environment: Scientific Issues*, ed. H. O. Halvorson, D. Pramer, and M. Rogul (Washington, DC: American Society for Microbiology, 1985), 23–35.

5. S. E. Lindow, "Methods of Preventing Frost Injury Caused by Epiphytic Ice-Nucleation-Active Bacteria," *Plant Disease* 67 (March 1983): 327–33.

6. Susan S. Hirano and Christen D. Upper, "Ecology and Physiology of *Pseudomonas syringae*," *Bio/Technology* 3 (1985): 1073–78.

7. S. E. Lindow, "Population Dynamics of Epiphytic Ice Nucleation Active Bacteria on Frost Sensitive Plants and Frost Control by Means of Antagonistic Bacteria," in *Plant Cold Hardness and Freezing Stress*, ed. A. Sakai and P. H. Li (New York: Academic Press, 1982), 395–416.

8. Lindow, "Methods of Preventing Frost Injury."

9. Lindow, "Population Dynamics."

10. Hirano and Upper.

11. Advanced Genetic Sciences, "Proposal to Field Test Genetically Engineered *Pseudomonas* Strains Containing Artificially Introduced Deletions in Ice-Nucleation Genes," submitted to the National Institutes of Health, Recombinant DNA Advisory Committee, 22 March 1984.

12. F. E. Young and H. I. Miller, " 'Old' Biotechnology to 'New' Biotechnology: Continuum of Disjunction," in *Safety Assurance for Environmental Introductions of Genetically-Engineered Organisms*, ed. Joseph Fiksel and Vincent T. Covello (New York: Springer-Verlag, 1987).

13. Donald S. Fredrickson, "The Recombinant DNA Controversy: The NIH Viewpoint," in *The Gene Splicing Wars*, ed. Raymond A. Zilinskas and Burke K. Zimmerman (New York: MacMillan, 1986), 15, 25.

14. Elizabeth Milewski and Bernard Talbot, "Proposals Involving Field Testing of Recombinant DNA-Containing Organisms," *Recombinant DNA Technical Bulletin* 6 (December 1983): 141–45.

15. Lindow, "Methods of Preventing Frost Injury," 332.

16. Advanced Genetic Sciences.

17. Lindow, "Ecology of *Pseudomonas syringae*."

18. National Institutes of Health, "Guidelines for Research Involving Recombinant DNA Molecules," *Federal Register* 48 (June 1983): 24580, appendix.

19. Frances L. McChesney and Reid G. Adler, "Biotechnology Released from the Lab: The Environmental Regulatory Framework," *Environmental Law Reporter* 13 (November 1983): 10366–80.

20. Ibid., 10368.

21. Stephen Jay Gould, "On the Origins of Specious Critics," *Discover* (January 1985): 34.

22. Leonard A. Cole, *Clouds of Secrecy* (Totowa, NJ: Rowman & Littlefield, 1988), 145.

23. Wayne Biddle, "Judge Forbids Army to Build Germ War Facility," *New York Times*, 1 June 1985.

24. Quoted from "Army Scales Down Dugway Lab," *GeneWatch* 5 (July-October 1988): 4.

25. Editorial, "A Novel Strain of Recklessness," *New York Times*, 6 April 1986.

26. Calestous Juma, *The Gene Hunters* (Princeton, NJ: Princeton University Press, 1989), 132.

27. Steven Schatzow, "The Role of the Environmental Protection Agency," *Genetically Altered Viruses and the Environment*, ed. Bernard Fields, Malcolm A. Martin, and Daphne Kamely (Cold Spring Harbor, NY: Cold Spring Harbor Laboratory, 1985), 50.

28. Ibid., 51.

29. Ibid., 56.

30. Amy S. Rispin, director, Science Integration Staff, Hazard Evaluation Division, EPA, *Memorandum*, to Tom Ellwanger, Registration Division, Office of Pesticides and Toxic Substances, 5 November 1985.

31. Ibid.

32. Eugene P. Odum, "Biotechnology and the Biosphere," *Science* 229 (27 September 1985): 1338.

33. Lindow, "Methods of Preventing Frost Injury," 332.

34. Julianne Lindemann, Gareth J. Warren, and Trevor V. Suslow, "Ice-Nucleating Bacteria," *Science* 231 (7 February 1986): 536.

35. Glen Church et al., letter to Monterey County Board of Supervisors, 3 January 1986.

36. National Working Group on Gene Technology, West German Green Party, telegram to Sam Karas, chairman, Monterey County Board of Supervisors, 15 January 1986. The telegram was signed by 27 members of the Green Party.

37. Members of the European Parliament, Strasbourg, France, telegram to Sam Karas, chairman of the Monterey County Board of Supervisors, 17 January 1986.

38. Sheldon Krimsky and Alonzo Plough, "The Release of Genetically Engineered Organisms into the Environment: The Case of Ice Minus," chapter 3 in *Environmental Hazards: Communicating Risks as a Social Process* (Dover, MA: Auburn House, 1988), 75–129.

CHAPTER 8

1. Frances E. Sharples, "Regulation of Products from Biotechnology," *Science*, 235 (13 March 1987): 1332.

2. Bernard Davis, "Bacterial Domestication: Underlying Assumptions," *Science* 235 (13 March 1987): 1329.

3. Thomas Kuhn, *The Structure of Scientific Revolutions* (Chicago: University of Chicago Press, 1962).

4. S. O. Funtowicz and J. R. Ravetz, "Global Environmental Issues and the Emergence of Second Order Science" (Presentation to the Commission of the European Communities, 1990).

5. Sheldon Krimsky, *Genetic Alchemy: The Social History of the Recombinant DNA Controversy* (Cambridge, MA: MIT Press, 1982).

6. Ernst Peter Fischer and Carol Lipson, *Thinking about Science: Max Delbruck and the Origins of Molecular Biology* (New York: W. W. Norton and Co., 1988).

7. Office of Technology Assessment, *Impacts of Applied Genetics*, OTA–HR–132 (Washington, D.C.: U.S. Government Printing Office, 1981), 240–54.

8. Sheldon Krimsky, "Patenting of Life Forms *sui Generis*: Some New Questions for Science, Law, and Society," *Recombinant DNA Technical Bulletin* 4 (April 1981): 11–15.

9. Peter Day, "Engineered Organisms in the Environment: A Perspective on the Problem," in *Engineered Organisms in the Environment: Scientific Issues*, ed. Harlyn O. Halvorson, David Pramer, and Marvin Rogul (Washington, DC: American Society for Microbiology, 1985), 9.

10. Philip Regal, "Biotechnology Jitters: Will They Blow Over?" *Biotechnology Education* 1 (1989): 51–55.

11. Ruth Hubbard, *The Politics of Women's Biology* (New Brunswick, NJ: Rutgers University Press, 1990), 65.

12. Quoted in Krimsky, *Genetic Alchemy*, 275.

13. Sharples, "Regulation of Products from Biotechnology," 1330.

14. Winston J. Brill, "Safety Concerns and Genetic Engineering in Agriculture," *Science* 227 (25 January 1985): 381–84.

15. Daniel L. Hartl, "Engineered Organisms in the Environment: Inferences from Population Genetics," in Halvorson, Pramer, and Rogul, eds., *Engineered Organisms*, 84.

16. Ibid.

17. Martin Alexander, "Ecological Consequences: Reducing the Uncertainties," *Issues in Science & Technology* 1 (Spring 1985): 60.

18. Philip J. Regal, "The Ecology of Evolution: Implications of the Individualistic Paradigm, in Halvorson, Pramer, and Rogul, eds., *Engineered Organisms*, 11–19.

19. Ibid., 12.

20. Ibid., 14.

21. Ibid., 19.

22. Sharples, "Regulation of Products from Biotechnology," 1330.

23. National Academy of Sciences, *Research with Recombinant DNA* (Washington, DC: National Academy of Sciences, 1977).

24. National Academy of Sciences, Committee on the Introduction of Genetically Engineered Organisms into the Environment, *Introduction of Recombinant DNA-Engineered Organisms into the Environment: Key issues* (Washington, DC: National Academy Press, 1987).

25. Ibid., 9.

26. Sheldon Krimsky, "Breaching Species Barriers," chapter 19 in *Genetic Alchemy*, 264–276.

27. National Academy of Sciences, *Introduction*, 21.

28. The issue of testability and its relationship to the empirical falsifiability of hypotheses is discussed in Karl R. Popper, *The Logic of Scientific Discovery* (New York: Harper and Row, 1959).

29. Frank E. Young and H. I. Miller, "Deliberate Releases in Europe: Over-Regulation May be the Biggest Threat of All," *Gene* 75 (30 January 1989): 1–2.

30. J. M. Tiedje et al., "The Planned Introduction of Genetically Engineered Organisms: Ecological Considerations and Recommendations," *Ecology* 70 (April 1989): 297–315.

31. Brill, "Safety Concerns."

32. Ibid., 383–84.

33. Nina V. Federoff, "Moving Genes in Maize," in Halvorson, Pramer, and Rogul, eds., *Engineered Organisms*, 75.

34. Day, "Engineered Organisms," 9.

35. Waclaw Szybalski [letter], *Science* 229 (12 July 1985): 112–13.

36. Elliott A. Norse, "Testimony on the Coordinated Framework for the Regulation of Biotechnology," Subcommittees on Investigations and Oversight; Natural Resources, Agricultural Research and Environment; and Science, Research and Technology, Committee on Science and Technology, 23 July 1986. Reprinted in *Recombinant DNA Technical Bulletin* 9 (September 1986): 171–77.

37. David Pimentel et al., *Environmental and Social Costs of Genetic Engineering* Unpublished Paper. (Ithaca, NY: College of Agriculture and Life Sciences, Cornell University, 1988).

38. Alexander, "Ecological Consequences," 61.

39. Pimentel, et al., *Environmental and Social Costs,* 6.

40. Philip J. Regal, "Splicing Genes Safely and Effectively," manuscript.

41. Office of Technology Assessment, *New Developments in Biotechnology—Field Testing Engineered Organisms: Genetic and Ecological Issues*, OTA–BA–350 (Washington, DC: U.S. Government Printing Office, 1988), 110.

42. Tiedje et al., "Planned Introduction," 306.

43. Eugene P. Odum, professor of ecology, University of Georgia, affidavit, 8 Sep-

tember 1983. Submitted with lawsuit filed by the Foundation on Economic Trends, 14 September 1983.

44. Robert K. Colwell et al., "Genetic Engineering in Agriculture," *Science* 229 (12 July 1985): 111–12.

45. G. Stotzky and H. Babich, "Fate of Genetically-Engineered Microbes in Natural Environments," *Recombinant DNA Technical Bulletin* 7 (December 1984): 163–88.

46. Halvorson, Pramer, and Rogul, eds., *Engineered Organisms*, 100–101.

47. Cited in ibid., 100.

48. Cited in ibid., 55.

49. National Research Council, *Field Testing Genetically Modified Organisms* (Washington, DC: National Academy Press, 1989), 2.

50. Office of Science and Technology Policy, "Coordinated Framework for Biotechnology Regulation," *Federal Register* 51 (November 1985): 47174–76.

51. Krimsky, "Breaching Species Barriers."

52. Sheldon Krimsky, "An International Affair," chapter 7 in *Genetic Alchemy*.

53. Sheldon Krimsky, "The Logic of the First NIH Guidelines," chapter 13 in *Genetic Alchemy*.

54. Davis, "Bacterial Domestication," 1334.

55. Hartl, "Engineered Organisms," 84.

56. R. K. Colwell et al., "Response to the Office of Science and Technology Policy Notice 'Coordinated Framework for the Regulation of Biotechnology,' " *Bulletin of the Ecological Society of America* 68 (1987): 16–23.

57. Sharples, "Regulation of Products," 1332.

58. Colwell et al., "Response," 21.

59. Kuhn, *Structure of Scientific Revolutions*, x.

60. Ibid.

61. Ibid., 150.

62. P. A. Schilpp, ed., *Albert Einstein: Philosopher-Scientist* (New York: Harper, 1959), 224–30.

CHAPTER 9

1. Martin J. Cline, "Insertion of New Genes into Living Organisms," *Proceedings, 1981 Battelle Conference on Genetic Engineering*, vol. 2 (Seattle, WA: Battelle Seminars and Studies Program, 1981), 245.

2. James D. Watson, "First Word," *Omni* 12 (June 1990): 6.

3. Susan Wright, ed., *Preventing a Biological Arms Race* (Cambridge, MA: MIT Press, 1990).

4. The Biological Weapons Anti-Terrorism Act of 1990 (P.L. 101–298) states that anyone "who knowingly develops, produces, stockpiles, transfers, acquires, retains, or possesses any biological agent, toxin, or delivery system for use as a weapon, or knowingly assists a foreign state or any organization to do so, shall be fined under this title or imprisoned for life or any term of years, or both."

5. Ethan Signer describes a proposal at a scientific meeting to replace the term "genetic engineering," which has a negative connotation, with the term "gene therapy." Ethan Signer, "Gene Manipulation: Progress and Prospects," in *Macromolecules Regulating Growth and Development*, proceedings of the 3d symposium of the Society for Developmental Biology (New York: Academic Press, 1974).

6. For a history of the eugenics movement see Daniel J. Kevles, *In the Name of Eugenics* (Berkeley: University of California Press, 1985).

7. Joshua Lederberg, "Experimental Genetics and Human Evolution," *American Naturalist* 100 (September/October 1966): 530. See also idem, "Genetic Engineering and the Amelioration of Genetic Defect," *Bioscience* 20 (December 1970): 1307—10; J. D. Watson, "Moving toward the Clonal Man: An Example of Scientific Inevitability?" *Atlantic Monthly* 227 (May 1971): 52–53.

8. Robert L. Sinsheimer, "The Prospect for Designed Genetic Change," *American Scientist* 57 (January-February 1969): 134–42.

9. Ibid.

10. Robert L. Sinsheimer, "The Presumptions of Science," *Daedalus* 107 (Spring 1978): 23–35.

11. Salvador Luria, "Modern Biology: A Terrifying Power," *Nation* (20 October 1969): 406–9.

12. Ibid., 409.

13. J. Beckwith, "Gene Expression in Bacteria and Some Concerns about the Misuse of Science," *Bacteriological Reviews* 34 (September 1970): 222–27.

14. J. K. Glassman, "Harvard Genetics Researcher Quits Science for Politics," *Science* 167 (13 February 1970): 963–64.

15. Theodore Friedmann and Richard Roblin, "Gene Therapy for Human Genetic Disease?" *Science* 175 (3 March 1972): 945–55.

16. June Goodfield, *Playing God* (New York: Random House, 1977), 52.

17. John Fletcher, testimony before the House Committee on Science and Technology, Subcommittee on Investigations and Oversight, 17 November 1982. *Human Genetic Engineering* (Washington, DC: U.S. Government Printing Office, 1983), 358. See also W. F. Anderson, "Genetic Therapy," in *The New Genetics and the Future of Man*, ed. M. P. Hamilton (Grand Rapids, MI: William B. Eerdmans, 1971), 117–18; Theodore Friedmann, *Gene Therapy: Fact and Fiction* (Cold Spring Harbor, NY: Cold Spring Harbor Laboratory, 1983).

18. Lederberg, "Genetic Engineering," 1309.

19. Signer, "Gene Manipulation," 227.

20. Sheldon Krimsky, *Genetic Alchemy: The Social History of the Recombinant DNA Controversy* (Cambridge, MA: MIT Press, 1982), 105.

21. Michael Rogers, *Biohazard* (New York: Alfred A. Knopf, 1977); Nicholas Wade, *The Ultimate Experiment: Man-Made Evolution* (New York: Walker & Co., 1977); June Goodfield, *Playing God* (New York: Random House, 1977).

22. Ted Howard and Jeremy Rifkin, *Who Should Play God?* (New York: Dell, 1977), 43.

23. Ibid., 42.

24. The techniques for these experiments are described in Karen E. Mercola and Martin J. Cline, "The Potentials of Inserting New Genetic Information," *New England Journal of Medicine* 303 (27 November 1980): 1297–1300.

25. Chairman, National Institutes of Health Ad Hoc Committee on the UCLA Report Concerning Certain Research Activities of Dr. Martin J. Cline, *Memorandum*, 21 May 1981, 8.

26. Ibid., 16.

27. Ibid.

28. Friedmann, *Gene Therapy*.

29. Ibid., 37.

30. Ibid., 59.

31. Ibid., 59–60.

32. Office of Technology Assessment, *New Developments in Biotechnology—Background Paper: Public Perceptions of Biotechnology—Special Report 2*, OTA–BP–BA–45 (Washington, DC: U.S. Government Printing Office, 1987).

33. Krimsky, *Genetic Alchemy*, 265.

34. See, for example, James D. Watson, *Molecular Biology of the Gene* (Menlo Park, CA: W. A. Benjamin, 1976).

35. A detailed analysis of the ice minus case is given in Sheldon Krimsky and Alonzo Plough, *Environmental Hazards: Communicating Risks as a Social Process* (Dover, MA: Auburn House, 1988), chapter 3.

36. The letter from the religious leaders is reprinted in President's Commission for the Study of Ethical Problems in Medicine and Biomedical and Behavioral Research, *Splicing Life* (Washington, DC: U.S. Government Printing Office, 1982).

37. Nicholas Wade, "Whether to Make Perfect Humans," *New York Times*, 22 July 1982, A22.

38. President's Commission for the Study of Ethical Problems in Medicine and Biomedical and Behavioral Research, *Splicing Life*, 3.

39. Ibid., 58.

40. Harold M. Schmeck, Jr., "Activities of Genes Reported Altered in Treating Man," *New York Times*, 9 December 1982, A1.

41. Colin Norman, "Clerics Urge Ban on Altering Germline Cells," *Science*, 220 (24 June 1983): 1360–61.

42. Eve K. Nichols, *Human Gene Therapy* (Cambridge, MA: Harvard University Press, 1988).

43. Ibid., 10.

44. Ibid.

45. Zsolt Harsanyi and Richard Hutton, *Genetic Prophecy: Beyond the Double Helix* (New York: Rawson, Wade, 1981), 152.

46. Elizabeth Milewski, "Development of a Points to Consider Document for Human Somatic Cell Gene Therapy," *Recombinant DNA Technical Bulletin* 8 (December 1985): 176–80.

47. National Institutes of Health, Recombinant DNA Advisory Committee Working Group on Human Gene Therapy (hereinafter cited as Working Group), "Points to Consider on the Design and Submission of Human Somatic Cell Gene Therapy Protocols," *Recombinant DNA Technical Bulletin* 8 (December 1985): 181–86.

48. National Institutes of Health, "Recombinant DNA Research: Proposed Action under Guidelines," *Federal Register* 51 (June 1986): 23210–11.

49. Ibid., 23211.

50. Ibid.

51. Working Group, *Memorandum*, 8 August 1986.

52. LeRoy Walters, "The Ethics of Human Gene Therapy," *Nature* 320 (20 March 1986): 227.

53. Ibid.

54. Bob Williamson, "Gene Therapy," *Nature* 298 (29 July 1982): 416–18.

55. Working Group, "Points to Consider."

56. Barbara J. Culliton, "NIH Asked to Tighten Gene Therapy Rules," *Science* 233 (26 September 1986): 1378–79.

57. Bernard Davis, "Cells and Souls," *New York Times*, 28 June 1983, A27.

58. Anderson, "Genetic Therapy," 109–24. W. F. Anderson, "Prospects for Human Gene Therapy," *Science* 226 (26 October 1984): 401–9; idem, "Human Gene Therapy: Scientific and Ethical Considerations," *Journal of Medicine and Philosophy* 10 (August 1985): 275–91; idem, "Human Gene Therapy: Why Draw a Line," *Journal of Medicine and Philosophy* 14 (December 1989): 681–693.

59. David Suzuki and Peter Knudtson, *Genethics* (Cambridge: Harvard University Press, 1989), 51.

60. Julie L. Gage, "Government Regulation of Human Gene Therapy," *Jurimetrics Journal* 27 (Winter 1987): 201–18.

61. Harsanyi and Hutton, *Genetic Prophecy*, 262.

62. OTA, *New Developments in Biotechnology*.

63. Ibid., 71.

64. Ibid., 72.

65. Ibid., 74.

66. This argument was developed in Sheldon Krimsky, "Must We Know Where to Stop before We Start?" *Human Gene Therapy* 1 (Summer 1990): 171–73.

CHAPTER 10

1. James J. Florio (D-NJ), "Regulation in Biotechnology," in *Biotechnology: Implications for Public Policy*, ed. Sandra Panem (Washington, DC: Brookings Institution, 1985), 44.

2. Winston J. Brill, "Why Engineered Organisms Are Safe," *Issues in Science and Technology* 4 (Spring 1988): 44.

3. David W. Fanning, "Issues Raised by Biotechnology: A Keystone Biotechnology Discussion Paper," presented at a conference on 14 July 1988. Keystone, CO: The Keystone Center.

4. Office of Technology Assessment, *Commercial Biotechnology* (Washington, DC: U.S. Government Printing Office, 1984), 355.

5. Ibid., 375.

6. Environmental Protection Agency, "Regulation of 'Biorational' Pesticides; Policy Statement and Notice of Availability of Background Document," *Federal Register* 44 (May 1979): 28093–95.

7. Environmental Protection Agency, *Pesticide Assessment Guidelines*. Subdivision M., Biorational Pesticides. Fred Betz et al. Office of Pesticides and Toxic Substances, EPA 540/9–82–028 (Washington, DC: Environmental Protection Agency, 1982).

8. Fred Betz, Simon Levin, and M. Rogul, "Safety Aspects of Genetically-Engineered Microbial Pesticides," *Recombinant DNA Technical Bulletin* 6 (December 1983): 135–41.

9. Sheldon Krimsky, *Genetic Alchemy: The Social History of the Recombinant DNA Controversy* (Cambridge: MA: MIT Press, 1982), 312–37.

10. Douglas M. Costle, letter to Adlai E. Stevenson, chairman, Senate Subcommittee on Science, Technology, and Space, 9 December 1977. Reprinted in Senate Committee on Commerce, Science, and Transportation, Subcommittee on Science, Technology and

Space, *Recombinant DNA Research and Its Applications* (Washington, DC: U.S. Government Printing Office, 1978), 88.

11. Ibid., 49.

12. Stanley H. Abramson, associate general counsel, Pesticides and Toxic Substances Division, memorandum to John A. Todhunter, Assistant Administrator for Pesticides and Toxic Substances, 14 March 1983.

13. Administrator's Toxic Substances Advisory Committee, *Memorandum. ATSAC Observations and Recommendations on Biotechnology*, 28 June 1983.

14. Don R. Clay, acting assistant administrator, EPA Office of Pesticides and Toxic Substances, testimony before the House Committee on Science and Technology, Subcommittees on Science, Research and Technology and Investigations and Oversight, 22 June 1983.

15. Environmental Protection Agency, Office of Toxic Substances, "Points to Consider in the Preparation of TSCA Premanufacture Notices for Genetically Engineered Microorganisms" (Draft document, 20 September 1984).

16. Environmental Protection Agency, Office of Toxic Substances, "Regulation of Research and Development, Field Releases of Microorganisms under TSCA," 20 September 1989, 12.

17. House Committee on Science and Technology, Subcommittee on Investigations and Oversight, *The Environmental Implications of Genetic Engineering*, staff report (Washington, DC: U.S. Government Printing Office, 1984).

18. Thomas O. McGarity, "Regulating Biotechnology," *Issues in Science and Technology* 1 (Spring 1985): 40–56. Gary Marchant, "Modified Rules for Modified Bugs: Balancing Safety and Efficiency in the Regulation of Deliberate Release of Genetically Engineered Organisms," *Harvard Journal of Law and Technology* 1 (Spring 1988): 163–208.

19. Environmental Law Institute, "TSCA: Overview of Statutory Authority and Regulatory Implementation" (Manuscript prepared by L. S. Ritts, 1984).

20. Marchant, "Modified Rules for Modified Bugs," 187–88.

21. Office of Science and Technology Policy, Executive Office of the President, "Proposal for a Coordinated Framework for Regulation of Biotechnology," *Federal Register* 49 (December 1984): 50857.

22. Ibid.

23. Ibid.

24. Ibid., 50904.

25. Ibid., 50858.

26. George A. Keyworth, memorandum to Federal Coordinating Council on Science, Engineering and Technology (FCCSET), 30 October 1985. Office of Science and Technology Policy, "Coordinated Framework for Regulation of Biotechnology; Establishment of the Biotechnology Science Coordinating Committee," *Federal Register* 50 (November 1985): 47174–76.

27. Office of Science and Technology Policy, "Coordinated Framework for Regulation of Biotechnology," *Federal Register* 51 (June 1986): 23309–93.

28. David Kingsbury, "Current Issues in Biotechnology Regulation: Government Perspective" (Presentation before the New England Biotechnology Association, Worcester Polytechnic Institute, 19 May 1987).

29. Environmental Protection Agency, Office of Pesticides and Toxic Substances,

Regulation of Research and Development, Field Releases of Microorganisms under TSCA,
20 September 1989, 12.

30. IGB Products brochure, circa 1986.

31. Mark Crawford, "NSF Official's Finances Probed by Justice," *Science* 238 (23
October 1987): 478; idem, "Document Links NSF Official to Biotech Firm," *Science*
238 (6 November 1987): 742; idem "Kingsbury Resigns from NSF," *Science* 242 (7
October 1988): 28.

32. Elizabeth Thompson, "Local Man Develops Gene-cloning Kit for Kids," *Patriot
Ledger*, 18 November 1985. Anne Michaud, "Fear Follows Collapse of Gene Research
Home," *Patriot Ledger*, 20 August 1987; idem, "Home Lab Accident Stirs Talk of
Tighter DNA Work Rules," *Patriot Ledger*, 26 August 1987; idem, "House-Collapse
Inquiry Stymied," *Patriot Ledger*, 21 August 1987.

33. Colleen Cordes, "Science, Politics Mix in Brouhaha over Field Test of New Pig
Vaccine," *Chronicle of Higher Education* 33 (8 October 1986): 1.

34. House Subcommittee on Investigations and Oversight, Committee on Science and
Technology, Subcommittee on Department Operations, Research, and Foreign Agriculture, Committee on Agriculture, *USDA Licensing of a Genetically Altered Veterinary
Vaccine*, joint hearing, 29 April 1986 (Washington, DC: U.S. Government Printing Office
1986), 306.

35. Ibid., 173.

36. Ibid., 291.

37. Ibid., 113.

38. Keith Schneider, "Gene Scientist Freed Germs in 1984 Tests," *New York Times*,
2 October 1987.

39. Peter W. Huber, "Biotechnology and the Regulation Hydra," *Technology Review*
90 (November/December 1987): 57–65.

40. The White House Council on Competitiveness, headed by Vice President Dan
Quayle, an office with the mandate to improve the U.S.'s global economic competitive
position, was given the task of drafting the long overdue scoping principles for regulating
biotechnology. The first draft was published in the *Federal Register* on July 31, 1990
under the aegis of the Office of Science and Technology Policy.

CHAPTER 11

1. Cary Fowler et al. "The Laws of Life," *Development Dialogue* nos. 1–2 (1988):
228.

2. Twenty-eight participants from 19 countries met at La Soleillette, Bogeve, France,
7–12 March 1987, for the Dag Hammarskjold Seminar on the Socioeconomic Impact of
New Biotechnologies on Basic Health and Agriculture in the Third World. A declaration
of goals for biotechnology, the Bogeve Declaration, was issued by the participants.
Fowler, "Laws of Life," 289–291.

3. Steven M. Gendel, "Biotechnology and Bioethics," in *Agricultural Bioethics*, ed.
S. M. Gendel et al. (Ames: Iowa State University Press, 1990), 342.

4. Ibid., 343.

5. Barry Commoner, "Bringing up Biotechnology," *Science for the People* 19
(March/April 1987): 12.

6. Jonathan King, "Patenting Modified Life Forms: The Case Against," *Environment*
24 (July/Aug 1982): 38.

7. Barry Commoner, *The Poverty of Power* (New York: Alfred A. Knopf, 1976).

8. Commoner, "Bringing up Biotechnology," 9.

9. Office of Technology Assessment, *Commercial Biotechnology: An International Analysis* (Washington, DC: U.S. Government Printing Office, 1984).

10. Office of Technology Assessment, *New Developments in Biotechnology—Field Testing Engineered Organisms: Genetic and Ecological Issues* (Washington, DC: U.S. Government Printing Office, 1988).

11. For a recent discussion of these ideas see Langdon Winner, *The Whale and the Reactor* (Chicago: University of Chicago Press, 1986).

12. For example, see Erwin Chargaff, "Profitable Wonders—A Few Thoughts on Nucleic Acid Research," *The Sciences* 15 (August/September 1975): 21.

13. In July 1990 it was reported that sex selection was being carried out by British physicians in preparation of embryo transplants. "Human Embryos Checked for Gene Defects," *Los Angeles Times*, 31 July 1990.

14. Cited in Gary Comstock, "The Case against BGH," in *Agricultural Bioethics*, ed. S. M. Gendel et al., 310.

15. Cary Fowler et al., "The Laws of Life: Another Development and the New Biotechnologies," *Development Dialogue* nos. 1–2 (1988): 143.

16. James B. Kliebenstein and Seung Y. Shin, "Impact of Bovine Somatotropin (BST) on Dairy Producers," in *Agricultural Bioethics*, ed. S. M. Gendel et al., 194.

17. Samuel Epstein, "Potential Public Health Hazards of Biosynthetic Milk Hormones," *International Journal of Health Services* 24 (January 1990): 78. Also, Samuel Epstein, "Growth Hormone Would Endanger Milk," *Los Angeles Times*, 27 July 1989.

18. Keith Schneider, "Stores Bar Milk Produced by Drugs," *New York Times*, 24 August 1989.

19. Robert J. Kalter et al., *Biotechnology and the Dairy Industry: Production Costs and the Commercial Potential of the Bovine Growth Hormone* (Ithaca, NY: Department of Agricultural Economics, Cornell University, 1984). Gary Comstock argues that even without BGH there is an attrition of dairy farms. He estimates the BGH contribution to that attrition at between 15.9 and 25.6 percent. Comstock, "The Case against BGH."

20. Jack Doyle, *Altered Harvest* (New York: Viking, 1985).

21. Foundation on Economic Trends, "Summary: Bovine Growth Hormone Issue" (Position paper, 1 April 1986).

22. Ibid.

23. Michael W. Fox, "Effects of Growth Hormone on Cows," *Science* 223 (19 September 1986): 1247.

24. Comstock, "The Case against BGH," 312.

25. Arthur H. Westing, ed., *Environmental Warfare* (London: Taylor & Francis, 1984); idem, *Herbicides in War: The Long Term Ecological and Human Consequences* (London: Taylor & Francis, 1984).

26. Calgene, *Annual Report*, Davis, CA, 1984.

27. Sheldon Krimsky, "Social Responsibility in an Age of Synthetic Biology," *Environment* 24 (July/August 1982): 2.

28. Sheldon Krimsky, "Biotechnology and Unnatural Selection: The Social Control of Genes," in *Technology and Social Change*, ed. Gene F. Summers (Boulder, CO: Westview, 1983), 51–70.

29. Doyle, *Altered Harvest*, 214.

30. Barry Meier, "Quest for Pest-Resistant Crops Is Raising Ecological Concerns," *Wall Street Journal*, 30 August 1985, 17.

31. Guido Ruivenkamp, "The Introduction of Biotechnology into the Pesticide Industry and Its Economic and Political Impact" (Working paper of the Vakgroep Internationale Betrekkingen en Volkenrecht, University of Amsterdam, 1985).

32. Rebecca Goldburg et al., *Biotechnology's Bitter Harvest*, a report of the Biotechnology Working Group, March 1990.

33. John Elkington, *The Gene Factory* (New York: Carroll & Graf, 1985), 83.

34. J. M. Tanner et al., "Effect of Human Growth Hormone Treatment for 1 to 7 Years on Growth of 100 Children with Growth Hormone Deficiency, Low Birthweight, Inherited Smallness, Turner's Syndrome, and Other Complaints," *Archives of Diseases in Childhood* 46 (1971): 745–82.

35. Guy Van Vliet et al., "Growth Hormone Treatment for Short Stature," *New England Journal of Medicine* 309 (27 October 1983): 1016–22.

36. Ibid., 1022.

37. Ross Feldberg, "Framing the Issue: The Social Implications of Biotechnology," *Science for the People* 17 (May/June 1985): 4.

38. Ibid., 9.

39. Gina Kolata, "New Growth Industry in Human Growth Hormone?" *Science* 234 (3 October 1986): 22–24.

40. Susan Okie, "Growth Hormone Reversed Aging in Study," *Boston Globe*, 5 July 1990, 3. See also Natalie Angier, "Human Growth Hormone Reverses Effects of Aging," *New York Times*, 5 July 1990, 1.

41. Samuel Epstein, *The Politics of Cancer* (San Francisco: Sierra Club Books, 1978).

42. Edward Yoxen, "Life as a Productive Force: Capitalising the Science and Technology of Molecular Biology," in *Science, Technology and the Labour Process: Marxist Studies*, ed. L. Levidow and R. M. Young, (London: CSE Books, 1981). S. Yanchinski, *Setting Genes to Work: The Industrial Era of Biotechnology* (Harmondsworth. Eng.: Penguin, 1985).

43. David Goodman, Bernardo Sorj, and John Wilkinson, *From Farming to Biotechnology* (Oxford: Basil Blackwell, 1987), 188.

44. Krimsky, "Biotechnology and Unnatural Selection," 68.

Selected Bibliography

Brunner, E. *Bovine Somatotropin: A Product in Search of a Market*. London: London Food Commission, 1988.

Busch, L. Lacey, W.B.; Burkhardt, J.; and Lacey, L.R. *Plants, Power, and Profit*. Oxford: Basil Blackwell, 1991.

Cooper, Iver P. *Biotechnology and the Law*. New York: Clark Boardman Co., 1989.

Dayan, A. D.; Campbell, P. N.; and Jukes, T. H., eds. *Hazards in Biotechnology: Real or Imaginary*. Proceedings of the Biological Council's symposium in London, 14–15 December 1987. New York: Elsevier, 1989.

Doyle, Jack. *Altered Harvest*. New York: Viking, 1985.

Elkington, J. *The Gene Factory: Inside the Biotechnology Business*. London: Carroll & Graf, 1985.

Fiksel, J., and Covello, V. T., eds. *Safety Assurance for Environmental Introductions of Genetically-Engineered Organisms*. New York: Springer-Verlag, 1988.

Gendel, Steven M.; Kline, David A.; Warren, D. Michael, and Yates, Faye, eds. *Agricultural Bioethics*. Ames: Iowa State University Press, 1990.

Goodman, D.; Sorj, B.; and Wilkinson, J. *From Farming to Biotechnology: A Theory of Agro-Industrial Development*. London: Basil Blackwell, 1987.

Hacking, Andrew J. *Economic Aspects of Biotechnology*. Cambridge: Cambridge University Press, 1987.

Hassebrook, Chuck, and Hegyes, Gabriel. *Choices for the Heartland: Alternative Directions in Biotechnology & Implications for Family Farming, Rural Communities & the Environment*. Ames: Iowa State University Press, 1989.

Jacobsson, S.; Jamison, A.; and Rothman, H., eds. *The Biotechnological Challenge*. Cambridge: Cambridge University Press, 1986.

Juma, Calestous. *The Gene Hunters: Biotechnology and the Scramble for Seeds*. Princeton: Princeton University Press, 1989.

Kenney, Martin. *Biotechnology: The University-Industrial Complex*. New Haven: Yale University Press, 1986.

Kloppenburg, Jack R., Jr. *First the Seed: The Political Economy of Plant Biotechnology*. Cambridge: Cambridge University Press, 1988.

Lappé, Marc. *Broken Code: The Exploitation of DNA*. San Francisco: Sierra Club Books, 1985.

Molnar, Joseph J., and Kinnucan, Henry, eds. *Biotechnology and the New Agricultural Revolution*. Boulder, CO: Westview Press, 1989.

National Research Council. *Field Testing Genetically Modified Organisms*. Washington, D.C.: National Academy Press, 1989.

Strauss, Harlee S. *Biotechnology Regulations: Environmental Release Compendium*. Washington, D.C.: OMEC International, 1987.

Suzuki, David, and Knudtson, Peter. *Genethics*. Cambridge, MA: Harvard University Press, 1989.

Teitelman, Robert. *Gene Dreams*. New York: Basic Books, 1989.

Woodman, William F.; Shelly, Mack C., II; and Reichel, Brin J. *Biotechnology and the Research Enterprise. A Guide to the Literature*. Ames: Iowa State University Press, 1989.

Wright, Susan, ed. *Preventing a Biological Arms Race*. Cambridge, MA: MIT Press, 1990.

Yoxen, Edward. *The Gene Business*. New York: Harper & Row, 1984.

Index

About the Author

SHELDON KRIMSKY is Professor of Urban and Environmental Policy at Tufts University. He is the author of *Genetic Alchemy: The Social History of the Recombinant DNA Controversy* and co-author of *Environmental Hazards: Communicating Risks as a Social Process.* He has published over 75 articles and reviews which have appeared in such distinguished publications as the *American Journal of Public Health, The Bulletin of the Atomic Scientists, Environment, Nature,* and *American Scientist.*